Editorial Board
编委会

星空间

星

间空

星空间

星艺装饰文化传媒中心　编著

暨南大學出版社
JINAN UNIVERSITY PRESS

中国·广州

图书在版编目（CIP）数据

星空间 / 星艺装饰文化传媒中心编著. —广州：暨南大学出版社，2017.10
ISBN 978 - 7 - 5668 - 2162 - 1

Ⅰ. ①星… Ⅱ. ①星… Ⅲ. ①建筑设计—作品集—中国—现代 Ⅳ. ①TU206

中国版本图书馆 CIP 数据核字（2017）第 191164 号

星空间
XING KONGJIAN
编著者：星艺装饰文化传媒中心

……………………………………………………………………………………………

出 版 人：徐义雄
策划编辑：胡艳晴　杜小陆
责任编辑：胡艳晴
责任校对：苏　洁
责任印制：汤慧君　周一丹

出版发行：暨南大学出版社（510630）
电　　话：总编室（8620）85221601
营销部（8620）85225284　85228291　85228292（邮购）
传　　真：（8620）85221583（办公室）　85223774（营销部）
网　　址：http: //www.jnupress.com
排　　版：广州良弓广告有限公司
印　　刷：深圳市新联美术印刷有限公司
开　　本：889mm×1194mm　1/12
印　　张：19.5
字　　数：235 千
版　　次：2017 年 10 月第 1 版
印　　次：2017 年 10 月第 1 次
定　　价：168.00 元

Happiness
and
Deliver Joy

设计幸福　播种快乐

星 艺 装 饰 文 化 传 媒 中 心

contents

chapter1：

第一章：公共·工程实景作品

chapter2：
第二章：住宅·工程实景作品

contents

chapter

1 公共·工程实景作品

FREE TIME 音乐咖啡吧

FREE TIME MUSIC COFFEE BAR

项目名称：FREE TIME 音乐咖啡吧
项目设计：广东星艺装饰集团
项目地址：广西桂林
设计师：龚志斌

本案业主是酷爱机车的小伙，几次沟通之后，设计师确定了本案设计风格：以现代手法，做出另类复古休闲的空间效果。对于一个商业空间来说，它要能吸引客源，要有市场竞争力。设计师调查了商业业态，经过不断的调整及完善，最终确定了本案的空间效果。整个空间色调统一，以灯光来烘托气氛，用软装饰丰富空间，加上造型元素贯穿整个空间，让客人来过之后都不会忘记这里。舒缓的音乐，泡上一杯咖啡，邀上恋人，下班后就是我们的 FREE TIME。

The owner of this scenario revels in locomotives. Thus after several communications with him, the designer determined the design style as using modern techniques to create a retrospective and leisure space effect. In the case of a commercial space, the design purpose is to attract customers and thus to build the market competitiveness. After investigating commercial activities and the continuous adjustment and improvement, the designer finally identified the spatial effect. The whole space is tonal and unified, which would leave guests a deep impression by using lamplight to foil atmosphere, using soft adornment to enrich the space, adding modeling element throughout the whole space. Try listening to the soothing music, having a cup of coffee with your beloved after work at FREE TIME.

LEARN
LIVE
HOPE

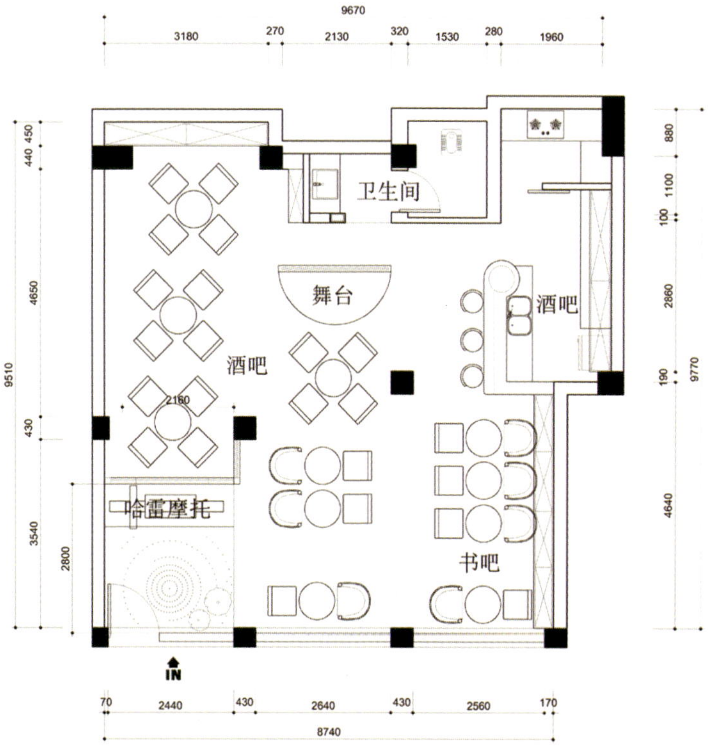
9670
3180
270
2130
320
1530
280
1960
卫生间
舞台
酒吧
酒吧
哈雷摩托
书吧
2160
440
450
4650
9510
430
3540
2800
880
1100
100
2860
190
9770
4640
IN
70
2440
430
2640
430
2560
170
8740

岑溪寿之元山茶油形象店

CENXI SHOUZHIYUAN CAMELLIA OIL IMAGE STORE

项目名称：岑溪寿之元山茶油形象店
项目设计：广东星艺装饰集团
项目地址：广西梧州
设计师：凌钦毅

本案是当地县城一个主营茶籽油的形象专卖店，设计理念主要突出自然、环保、生态。设计师用浅色木饰作为整体的基调，整个空间一进大门就可以看到一个很长的厨房岛台，背靠吊柜，营造出生活的气息，弱化了商店的感觉，让客户选购产品的时候感觉更加舒适。

This scenario is an image store in the local town, specialized in selling camellia oil. The natural flavor, environmental and ecological protection are highlighted in the concept of the design. Using tinted wood as the general tone, a long kitchen island, which is back against the wall cupboard, could be seen the moment you enter the space. The whole decoration creates a breath of life and thus weakens the store feeling, which makes customers feel more comfortable while purchasing goods.

Organic Camellia Oil
长寿之乡
优选品种
天然有机
低温冷榨
壽之元
NATURE LONG
有机冷榨山茶油
精选纯正岑溪软枝油茶

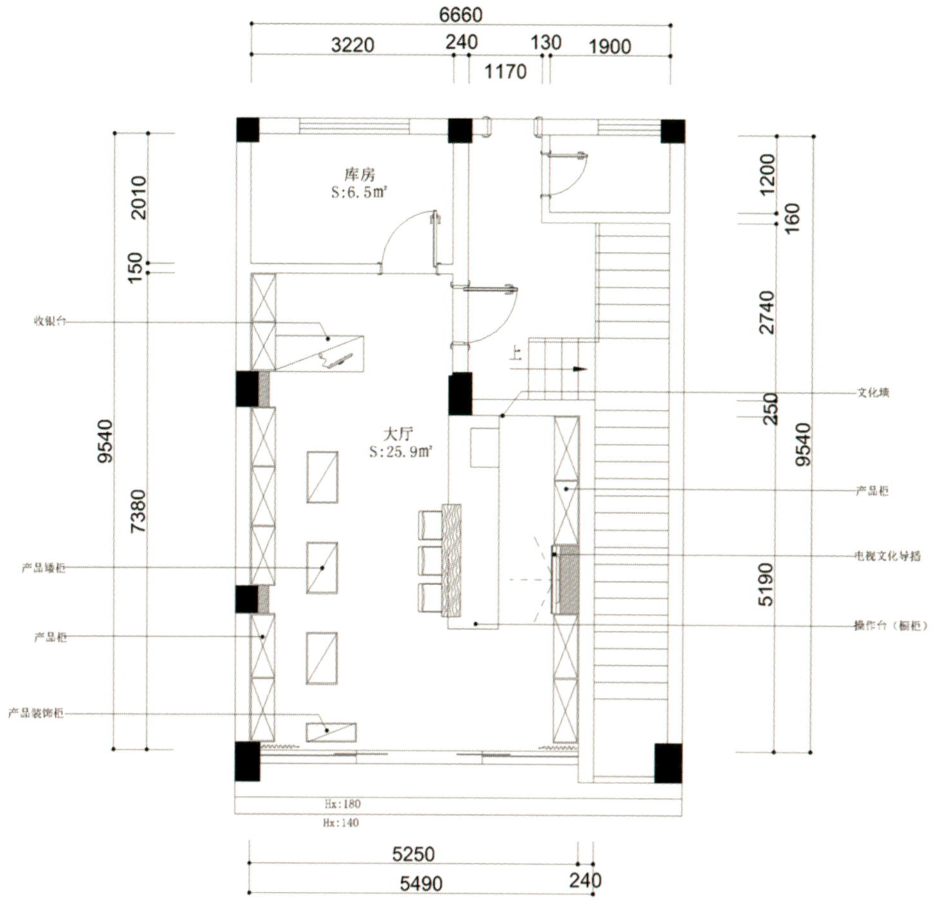
6660
3220
240
130
1900
1170
库房
S:6.5m²
大厅
S:25.9m²
上
2010
150
9540
7380
1200
160
2740
250
9540
5190
收银台
产品墙柜
产品柜
产品装饰柜
文化墙
产品柜
电视文化导语
操作台（橱柜）
Hx:180
Hx:140
5250
5490
240

新养道
“油”内而外
有机冷榨山茶油
暂停服务

茶之然

NATURAL-TEA

项目名称：茶之然
项目设计：广东星艺装饰集团
项目地址：广西南宁
设计师：韦莹莹

这是一家以茶为主题的工业风与中式混搭的创意生活馆，透露着一种云淡风轻的感觉，稳重大气，宁静淡泊，意境悠长。门外的桌椅、红砖、铁艺和饰品体现出工业风冷静、理性的特点。入门宽敞的吧台、灰色的墙板，迎合了中式家具讲究的内敛质朴的设计风格，体现了空间不一样的质感。设计师用非常纯粹的格子架展示茗茶与相关的物品，规矩的格子增加了空间立体感，也将空间的收纳做到最大化。店内台阶下错落的小空间相对独立地向店外展示着店内的商品，却又不失其整体性。楼梯处的照片墙展示着茶文化的历史，配上绿植墙、鸟笼饰品和实木铺设的阶梯，犹如穿过历史的间隙。

二楼的各色木质的沙发、椅子与桌子，体现出温润的感觉，有助于营造绝对的空间宁静感。没有墨守成规地采用统一的家具，不需要隔断，就可以依靠不同风格的家具与墙饰构造出独立而安静舒适的空间，体现了对家具布局细节的追求。

综观整个设计，结合了茶礼、茶器、茶食、茶饰、茶书的茶主题传统文化，含蓄内敛，体现出东方式的文化内涵和异域特色。设计师“追求时尚的同时，在喧嚣中独守一片宁静，还心灵一块净土”的设计理念，完美融合了传统文化与现代装饰。

This scenario is a tea-themed creative life pavilion, mixed with an industrial and Chinese style, which reveals a sense of sprezzatura — composed and formal, quiet indifferent, and a profound artistic conception. Outside of the door, the tables and chairs, red brick, iron art and accessories all embody the cool and rational characteristics of industrial style. The spacious bar table and the grey wallboard at the entrance comply with the restrained and simple Chinese style, which embodies a different feeling of the space. The lattice frame, which increases the sense of three-dimensional space as well as maximizes the storage space, is designed to display various teas and related items. Goods are displayed in the scattered small space under stairs, which is in integrity from outside. The stairwell wall of photographs decorated with plants, cage accessories and solid wood staircase show the history of tea culture, making people feel like wandering in the history.

The wooden sofas, chairs and tables in various colors in the second floor convey the warm feeling, which contributes to shaping the tranquil space. Instead of decorating with the unit furniture, the designer makes use of different styles of furniture and wall hangings to create an independent, sedate and comfortable space and thus in pursuit of furniture layout details.

From the perspective of the whole design, the reserved traditional tea culture, including tea ceremony, tea sets, tea accessories and tea books, embodies oriental culture connotations and exotic features. The design concept of finding a peaceful heart in bustle and hustle life as well as in pursuit of fashion is perfectly blended into traditional culture and modern decoration.

Nature / International

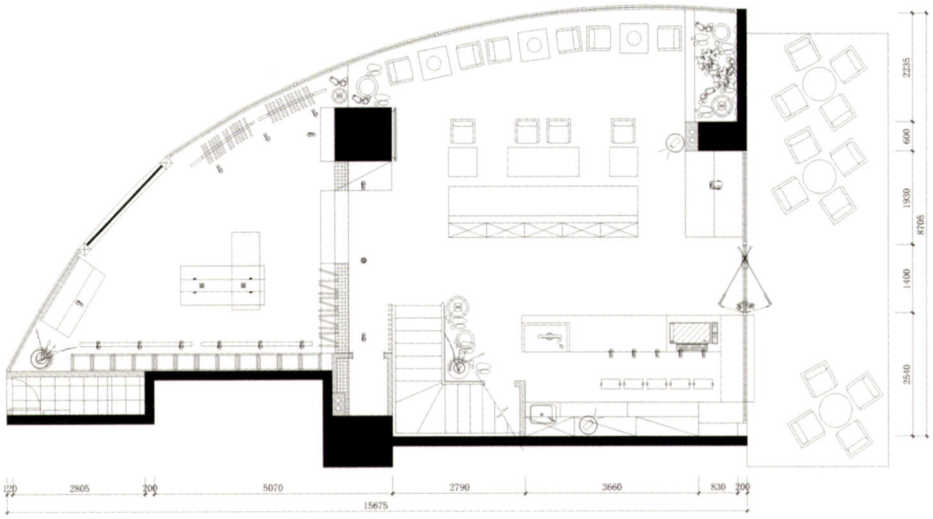
120
2805
200
5070
2790
3660
830
200
15675
2235
600
1930
1400
2540
8705

大化“因为所以”咖啡馆

DAHUA FINANCE COFFEE HOUSE

项目名称：大化“因为所以”咖啡馆
项目设计：广东星艺装饰集团
项目地址：广西南宁
设计师：覃雷　熊峰

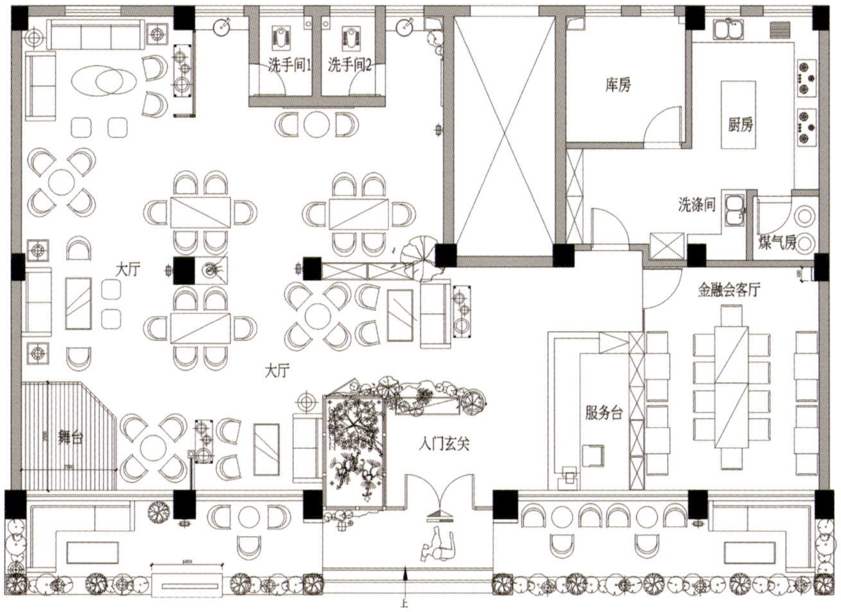

本案是一个彰显地域文化的壮族风格咖啡厅。橙色、红色、黄色和蓝色相互搭配，色彩十分浓烈，艳丽的色彩带来强烈的冲击力，体现了壮族人民的热情。极具民族特色的软、硬装一起营造了和谐的整体氛围，各类印花图案带来浓浓的民族风，丰富的层次感使整体空间非常耐看，有一种历久弥新的韵味。内部的材料虽普通，但奔放热情，气势不凡。摆入什么样的陈设就会随之呈现什么样的风格，有很强的融合感。外部的网格状花架与鲜花错落有致，浪漫温馨，昏暗的灯光氛围增加了神秘感，让人卸下工作包袱放松地享受生活。

This scenario is a coffee shop manifesting regional culture of Zhuang style. In the design, orange, red, yellow and blue are intertwined with each other. The bright colors bring out a strong impact, which embodies Zhuang people's enthusiasm. Superior ethnic characteristics of soft and hard decoration create a harmonious atmosphere. Various printing patterns reveal a strong national style. Rich layering style makes the space look very engaging, retaining a long-lasting appeal. Though the common materials are used, the unrestrained passion and extraordinary momentum are fully manifested. Various styles are shown by various layouts, which are fully integrated. The grid-shaped flower shelves and flowers are well-proportioned, which creates a romantic and cozy atmosphere. The dim light adds the mystery atmosphere, making people feel much more relaxed.

东篱田居土菜馆

DONGLI LOCAL RESTAURANT

项目名称：东篱田居土菜馆
项目设计：广东星艺装饰集团
项目地址：广西柳州
设计师：魏湛雨

福

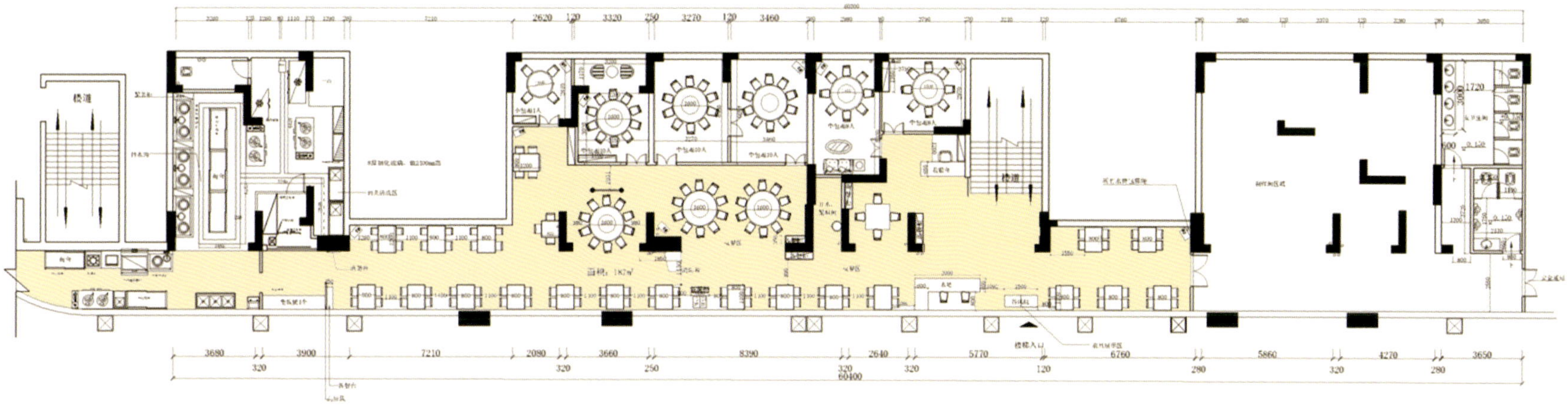

本案在大的方向上遵循中式设计中的对称格局，同时又有材质、色彩、大小、数量上富于变化的均衡形式，在一定程度上弥补了对称的单调与呆板。对称与均衡使复杂形态结构形成秩序性，展现出协调的形式美。

This scenario adheres to the Chinese symmetrical pattern, with material, color, size and number in variations, which compensates for the tedious symmetry in some degree. Meanwhile, the coordination of formal beauty is manifested in sophisticated structure of balance and symmetry.

菲诺时尚服装总部办公室

FINO FASHION CLOTHES HEADQUARTER OFFICE

项目名称：菲诺时尚服装总部办公室
项目设计：广东星艺装饰集团
项目地址：广西南宁
设计师：凌立成

本案为服装贸易类办公场所，设计师在设计之初确立了一种“对比与反差”的概念，使这个办公场所的设计无论是从整体氛围、空间划分还是材料的使用上，都能最大限度地凸显时尚感。

设计师认为，在南宁这个夏热冬暖的典型亚热带季风气候区，享受清凉大概是比较奢侈的事情了，所以设计师决定把办公场所设计成黑白分明的样子，来营造一个有“凉意”氛围的场所。当员工进入办公室，可以感受到一个完全不同的世界。从这种概念出发，空间的设计方案就一点点地拓展开来了。

首先外墙运用了点的元素，每一个圈都有自己的形态，这样一来，就跳脱了原始的想象，象征着员工无穷的潜力汇聚成一个面，努力向周围拓展。良好的寓意加上精巧的设计，将本案设计的精致典雅逐步呈现出来。办公区域的简洁明朗，显示出办公区的人性化与恰到好处，前台与茶水区的优美弧度，暗喻着这个空间的兼容并蓄。

在材料的使用方面，设计师更是考虑到运用最能体现空间概念的元素——不锈钢、黑玻、反复的白等，这些材料互相独立的同时又形成了一种奇妙的融合感。

This scenario is a clothing trade office. The concept of "comparison and contrast" was identified at first, which highlighted the site's fashion sense in the overall atmosphere, the spatial devision and material using.

Nanning is a typical subtropical zone，which is impossible to enjoy the cool and thus the designer decides to make it look like black and white to create a "chill" space. In this way, when the workers step into the office, they can feel cool. Based on such design concept, the design proposal is determined finally.

At first, point elements are applied to the exterior wall. Every circle of the wall has its own form which is an embodiment of convergence of workers' potential, striving to expand further. By utilizing such an excellent design concept and delicate design, the refined and elegant design of this scenario is manifested. The humanized working area is shown by the simple and clear design. The front desk and tea area are perfectly located, showing the tolerant atmosphere.

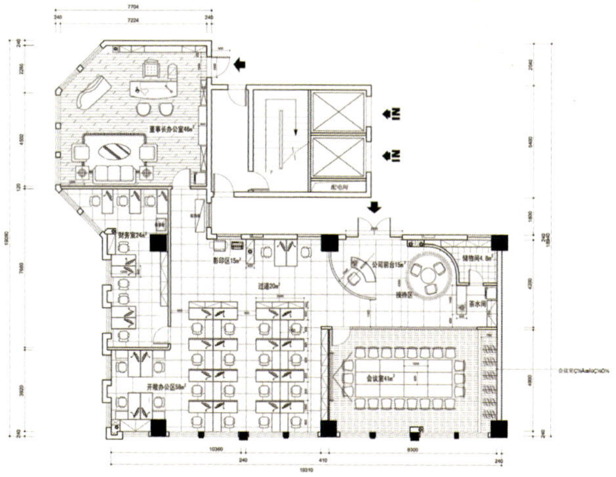

平面布置图 1:100

In terms of material, the stainless steel, black glass and white-colored material are independently used while at the same time they are integrated with each other harmoniously.

贵港明悦大酒店

GUIGANG MINGYUE HOTEL

项目名称：贵港明悦大酒店
项目设计：广东星艺装饰集团
项目地址：广西贵港
设计师：范建国　凌立成

设计初期，通过对地理环境、位置、客人消费能力、建筑面积大小、功能划分等多方面的调查，以及与客房的多次沟通，我们把酒店定位为达到四星级标准的高端商务酒店，采用现代简约的手法表现出中西文化韵味，使入住的客人能感受到高端大气且优雅舒适的艺术气息。

一进入大堂，整个空间以米黄色暖基调为主，并点缀以长条金色不锈钢造型，宽敞明亮又极具吸引力。具有中式祥云寓意的大理石地面拼花，倒映着椭圆形的层叠水晶大吊灯，沉稳大方，华丽优雅。大堂前台深色的木饰面、蛛网大理石以及抽象的装饰画，无不体现出整个酒店庄重沉稳又具有亲和力的灵魂思想，这也成为贯穿整个设计方案的主线。

自助餐厅摒弃了传统的中式风格，以屏风隔断为区域划分，以统一的深色木纹为主要装饰表现手法，搭配不规则拼贴的浅黄色墙砖，简单大方，使每一个用餐人员都能感到轻松自然。客房在满足数量要求的基础上变化房型，以满足商务客人以外的其他客人的需求。每间客房均按四星级标准配备设施，通过简洁的深色木边框、墙纸、布艺以及订制的成品家具等，营造出淡雅的东方韵味。卫生间的设计更给客人以想象空间——通透，又具有其私密性。

In early design stage, according to the investigation of various aspects, such as geographical environment, location, consumption ability, size of building, and functional partition and several communications with guest room, the hotel is positioned as high-end four star business hotel. Chinese and western culture is manifested by the use of modern minimalist approach, and thus the customers could experience the high-end and elegant artistic atmosphere.

Upon entering the lobby, the warm yellow lights, decorated with golden stainless steel strip are clearly seen, which makes the whole space seem spacious and attractive. What's more, the marble parquet with Chinese auspicious cloud reflects the oval crystal cascade chandelier and all the decoration makes the space seem graceful and elegant. The dark wood surface of the reception desk, the spider marble and the abstract drawings all reveal the dignified and friendly soul of the hotel, which is the main thread of the entire design as well.

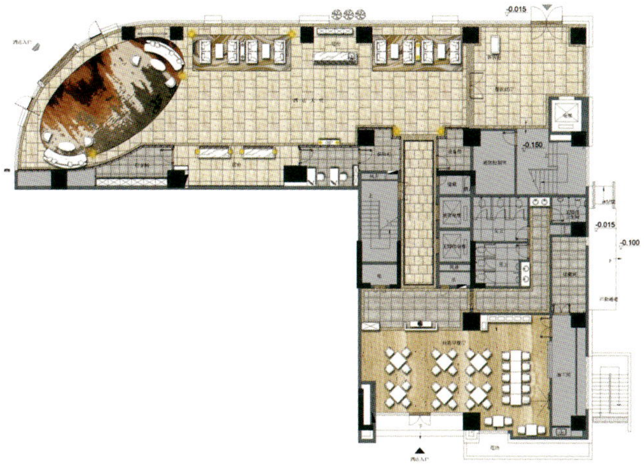

Without the traditional Chinese style, areas in the cafeteria are divided by screens. The cafeteria looks simple and decent with unified dark wood and irregular pale yellow wall bricks, which could make every customer feel at ease. Based on the room numbers, different rooms are decorated to meet the needs of various kinds of customers. Decorated with simple dark wood borders, wallpaper, fabric, and custom-made furniture, each room is equipped with four-star standard facilities, creating an oriental sense of elegance. The design of bathroom is full of imagination, enjoying transparency and privacy at the same time.

桂林香樟林大酒店

GUILIN FRAGRANT GARDEN VILLA HOTEL

项目名称：桂林香樟林大酒店
项目设计：广东星艺装饰集团
项目地址：广西桂林
设计师：范建国　许舰

本案通过提炼桂林这一著名旅游城市的特色元素，构造出富有传统地域文化内涵的现代精品酒店空间，通过“形”与“色”的组合转化，突出“形·色桂林”的主题概念。从整体入手，从使用功能出发，以人为本，在建筑的基础上深化、细化，使建筑体与周边环境相融合，一气呵成。一流的建筑，一流的设计，不在于昂贵材料的堆砌，而在于它的独特性、合理性、人文性。本案的设计宗旨为：实用美观，合理规划；以人为本，绿色环保。

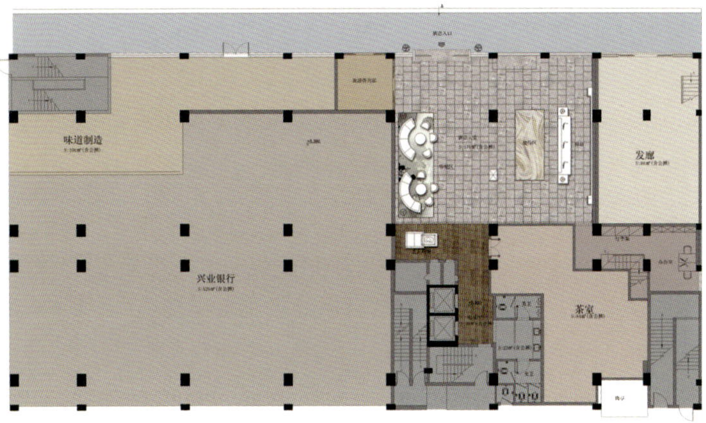
味道制造
发廊
兴业银行
茶室

This scenario aims to create a modern boutique hotel with traditional regional connotations by refining the element of the famous tourist city—Guilin. By the combination of “shape” and “color”, the concept of “Colorful Guilin” is highlighted. Holistically, taking account of utility and the principle of people-oriented, architecture is designed to be in harmony with the surrounding environment by-refining and deepening the foundation of it. The first-class building and design does not consist in its expansive materials but in its uniqueness, rationality and humanity. The design purpose of this scenario focuses on practicality and aesthetic, reasonable planning, people-oriented and environmental-friendly.

208
209

汉军 · 五象一号空中体验馆

HANJUN · WUXIANG NO . 1 AERIAL EXPERIENCE MUSEUM

项目名称：汉军 • 五象一号空中体验馆
项目设计：广东星艺装饰集团
项目地址：广西南宁
设计师：张至　郝建勋

本案位于广西壮族自治区南宁市邕江江畔，可观赏到一线江景。由于其优越的地理位置，最终被定位为以海员俱乐部为主题的空中体验馆。从登船区进来，映入眼帘的是一艘即将出海的木船，整个船体用防腐木和铁皮现场施工完成，让人身临其境——当你坐在船头甲板上，视线与窗外的江景接触的一刹那，会觉得这艘船正在起航，迎着江面那起伏不定的朵朵浪花缓缓前行，把汉军·五象一号的地理优势体现得淋漓尽致。地面、墙面及天花都没有做过多的修饰处理，大部分保留原有的水泥墙面，利用配饰、灯光效果来营造氛围。设计师在配饰方面也下足了功夫，如墙面壁画都是设计师亲自绘制，登船区的麻绳网也是设计师现场编制的。

This scenario is located in the riverside of Yong River, Nanning City, Guangxi Zhuang Autonomous Region with a full-view of the river. It is defined as the aerial experience museum with the seamen's club as the topic due to its superior geographic location. After entering into the boarding station, what jumps into your sight is a wooden boat which is about to be put out to sea. The whole boat is made of antiseptic wood and iron sheets under site operation, which gives you a feeling of actually being there. When you sit on the forecastle and your sight reaches the river view outside the window, you will feel this boat is setting sail and slowly moving ahead towards the rolling spray in the surface of the river, which gives full expression of the geographical advantage of Hanjun Wuxiang No.1 Ground. Wall and ceiling haven't been decorated too much, and most of them keep the original cement wall and people take advantage of accessories and lighting effects to create atmosphere. Designers also make lots of efforts in the aspect of accessories, for instance, murals in the wall are drawn by designers personally and hemp rope nets in boarding area are weaved by designers on spot.

WELCOME
ABOARD

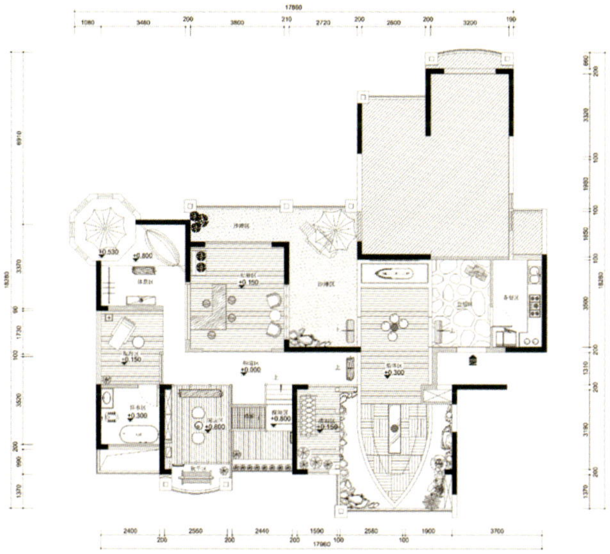

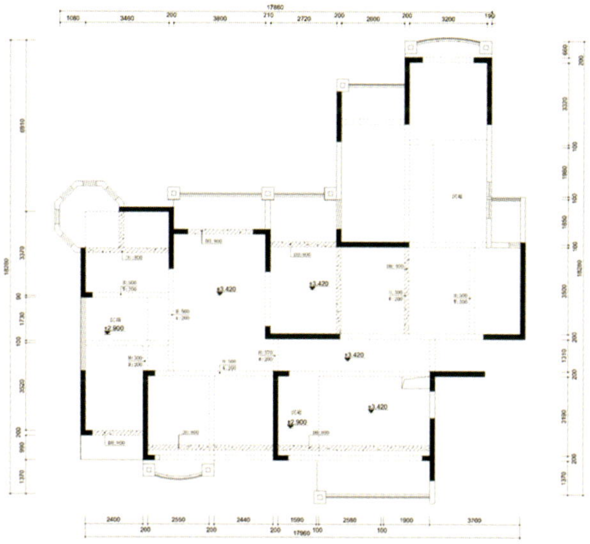

广西丽格假日酒店

GUANGXI LIGE HOLIDAY HOTEL

项目名称：广西丽格假日酒店
项目设计：广东星艺装饰集团
项目地址：广西南宁
设计团队：凌立成　凌钦毅　覃朝艺

丽格假日酒店
LIGE HOLIDAY HOTEL

本案位于繁华的青秀区会展中心附近，为了与商业圈整体构建相融合又不失新意，在前期设计方案中，让商务和艺术元素相辅相成，是设计师重点考虑的问题。

本案在满足基本功能的要求之余，在材料的设计上“大做文章”，大量使用地毯、跳色的皮革、极具设计感的家具和艺术品，以彰显其个性，力求打造在千变万化中仍无违和感的视觉体验。

将空间打造得既艺术又商务，不只承载了文化、生活的态度，更为住客创造了一个沉静心灵、安顿身心的悠闲空间。

This scenario is located in the neighborhood of bustling convention center of Qingxiu District. In order to integrate into the whole construction of business circle and not to lose the new meaning, in the previous design scheme, making business elements and art elements to supplement each other is the problem considered emphatically by designers.

Except to satisfy the requirements of basic functions, it makes a big fuss on the design of materials, namely, makes too much use of carpets, leather with different colors, furniture and artworks with high sense of design to display personality and strive to create sight experience with no feeling of strangeness in the myriads of changes.

Creating the space into an artistic and commercial one not only bear people's attitude on culture and life, but create a leisure space where lodgers are able to calm their heart and arrange for their mind and body.

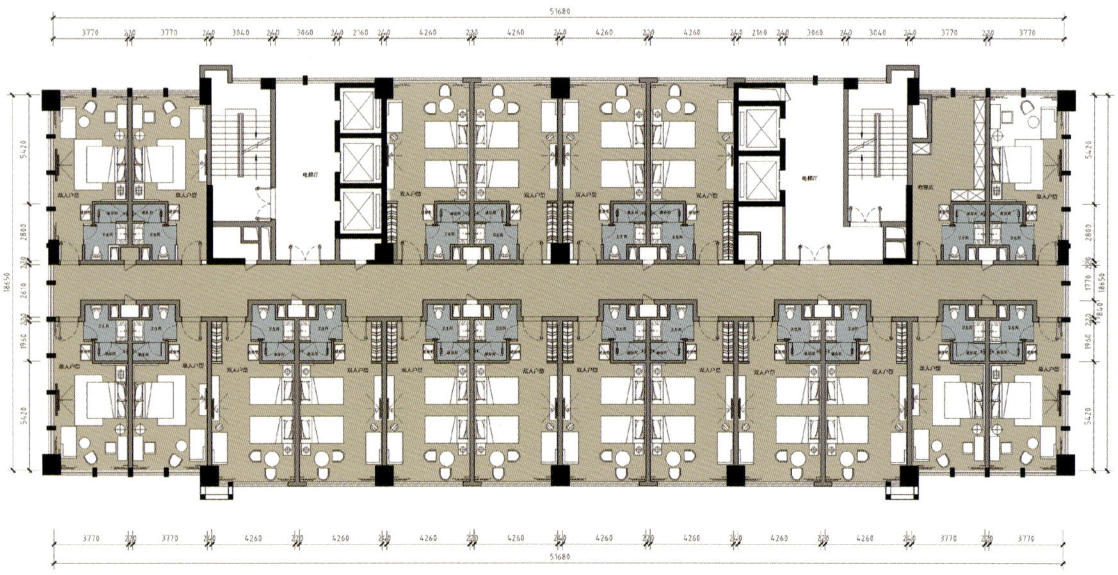

今日早餐

莉水堂休闲吧

LILY JUICE LOUNGE BAR

项目名称：莉水堂休闲吧
项目设计：广东星艺装饰集团
项目地址：广西百色
设计师：余涵龙

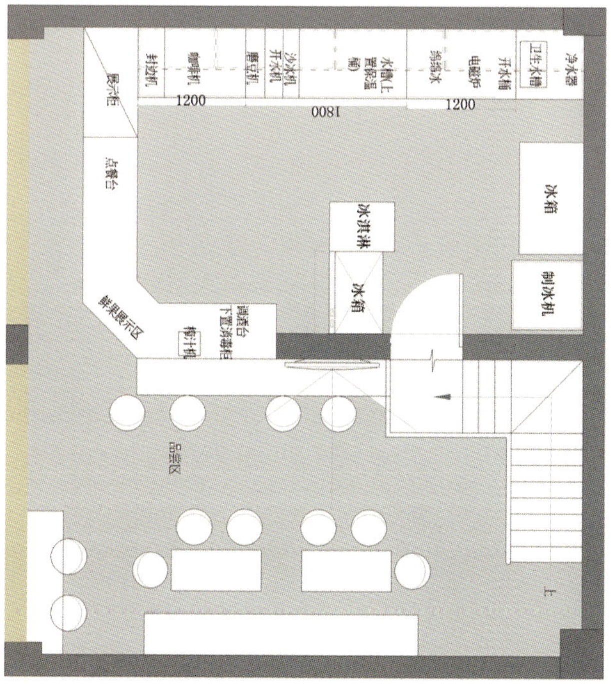

这是百色第一家loft风格的奶茶店，此店的面积只有40多平方米，设计师充分利用空间，采用双层设计结构，在寸土寸金的位置扩大了实用面积，多出的面积被设计成一个休闲书吧。原木、裸露的砖墙、工业风格的灯饰强化了设计感，营造出一种特别的闲适氛围。这样巧妙的设计为百色的夜晚增添了一个休闲的好去处。

This scenario is the first milk tea shop in loft style in Baise. With an area of more than 40 square meters, designers take full advantage of its space, adopt the design structure of double layers, expand the effective area in the high cost of land, as a result, the extra area is designed into a leisure book bar. Logs, bare brick wall and decorative lighting in the industrial style strengthen the sense of design and create a special leisure atmosphere.The ingenious design adds a good place to relax for the nights in Baise.

南宁蔓华·艺品轩文化沙龙

NANNING MANHUA · YIPINXUAN CULTURAL SALON

项目名称：南宁蔓华·艺品轩文化沙龙
项目设计：广东星艺装饰集团
项目地址：广西南宁
设计师：黄林妮

“林尽水源，便得一山，山有小口，仿佛若有光。”陶渊明《桃花源记》中的这一句，恰好能形容这间小店，幽静的店面设计使其与周围的高楼大厦相比仿佛桃花源。

从外观看就犹如步入古时候的小院一样，青石板铺设的“曲径”引人穿过庭院，在满园秋色中，黄色的木柜及上面摆放着的木质饰品突出一个“古”字。超高的层高配上落地的锦画，墙上的大幅挂饰锦荷花，再加上中式的家具，让人如同进入宫殿一般。

“Approaching the beginning of stream in the peach blossom land, he trudged until to a hill where a light seemingly dimmed in the depth.” The sentence from *A Story of the Peach Blossom Land* by Tao Yuanming, a renowned poet in ancient China, is precise to describe the little shop, for its designs of tranquility make the shop like the legendary peach blossom land by comparison with the surrounding skyscrapers.

Similar to strolling in the little yard in ancient times from its appearance, one tends to be attracted by the twisty path paved by green granite stone . In the middle of autumn, there stands a yellow wooden cabinet, on which there is a wooden ornament with an eye-catching Chinese character “Gu”. The large brocade painting with lotus flowers, hung from the lofty ceiling to the ground and against the Chinese furniture, makes one feel like walking into a Chinese palace.

二楼以红酒为主题，采用的是欧式设计。壁炉、沙发和精致的茶几，都带给人私人酒庄的体验。

三楼以花艺、茶为主题，大胆地采用了红蓝撞色，分成了包厢与大厅两个空间，古物的摆放、包厢前的青石板与鹅卵石、中式饰品和图案营造出纯正的东方情调，体现出价值感和文化气质，与一楼相互呼应。整体感观显现出空间的内涵，古朴典雅的器具透露着高贵的舒适感。壁饰则以中国风入题，营造出古典韵味。

Themed by red wines, the shop in the second floor is designed in European style. And fireplace, sofa as well as the exquisite tea table, bring one the experience of personal Chateau.

With the topic of floriculture and tea, the shop in the third floor alternates between boldly red and blue, falling into two parts as a box and a hall, in which the placed antiques, green granite stones and cobble stones in front of the box, as well as Chinese decorations and designs presented a type of oriental style. Thus the sense of value and culture in it go together with the atmosphere in the first floor. Its overall impression leaves visitors to learn about the spatial artistic conception and the quaint wares offer a noble and cozy conception. Apart from this, the accessories on the walls are themed by Chinese style, which present a classical charm.

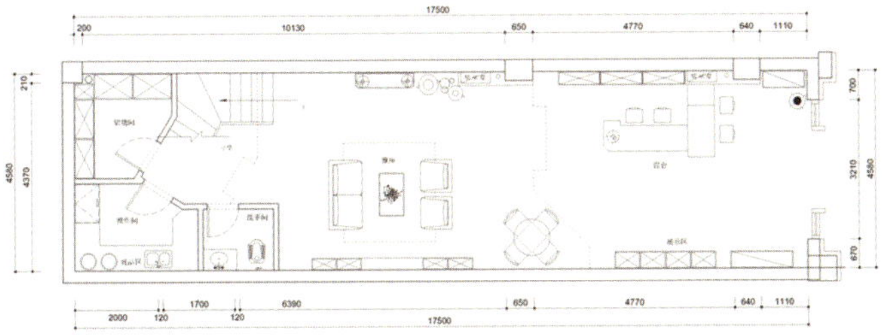
17500
200
10130
650
4770
640
1110
210
4580
4370
700
3210
4580
670
2000
120
1700
120
6390
650
4770
640
1110
17500

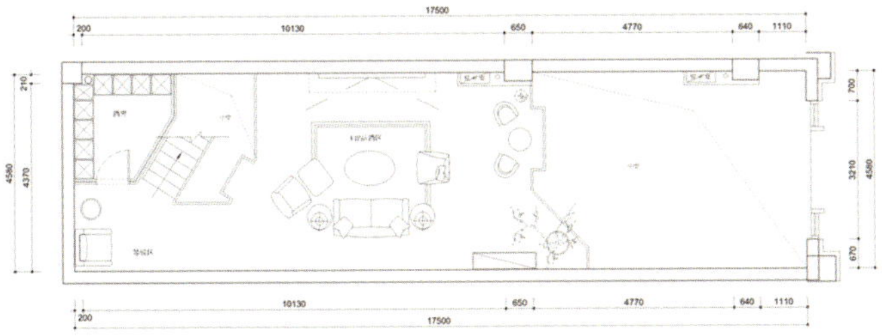
17500
200
10130
650
4770
640
1110
210
4580
4370
700
3210
4580
670
10130
650
4770
640
1110
200
17500

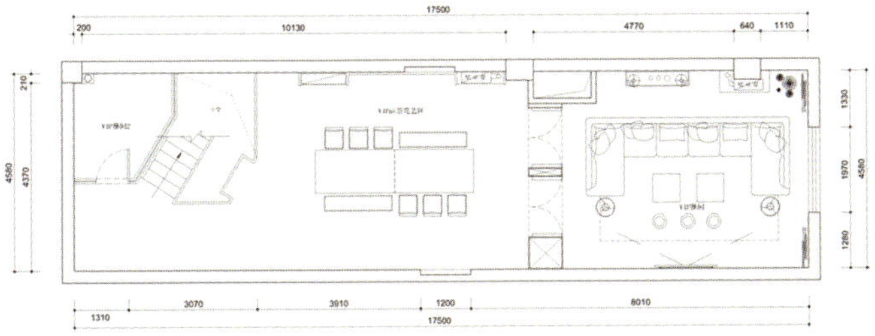
17500
200
10130
4770
640
1110
210
4580
4370
1330
1970
4580
1280
3070
3910
1200
8010
1310
17500

南宁思慕昔芒果冰吧

NANNING SMOOTHIE MANGO ICE BAR

项目名称：南宁思慕昔芒果冰吧
项目设计：广东星艺装饰集团
项目地址：广西南宁
设计师：覃雷　熊峰

设计师运用大量的木料与店外的钢筋水泥形成强烈的对比，通透的落地玻璃大方展现着内外的风光，简约的格局让错落的书籍显得格外丰富。本店充分利用有限空间，大块统一的颜色干净无比，随处可见的涂鸦与复古的饰品，都散发着小清新的韵味。

The designer applies masses of timber against the concrete jungle outside, with transparent French windows presenting the sight both in and out and with the simple layout manifesting the randomly strewn books plentiful. The shop makes most of the limited space, with coordinated large scaled colors looking like remarkably clean and wide range of drawing as well as quaint ornaments, all these emanating a beauty of brightness and cleanness.

¥42

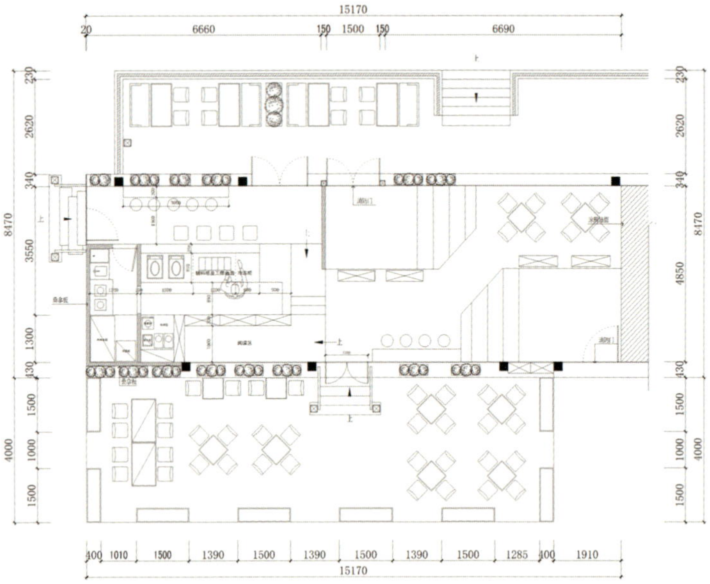
15170
6660
150
1500
150
6690
230
2620
340
3550
130
1300
1500
1000
4850
8470
4000
400
1010
1500
1390
1500
1390
1500
1390
1500
1285
400
1910
15170

Hush Puppies
ENJOY ONE'S LIFE
Welcome
HAPPY EVERY D
SUERTE

FUSE
Harlan COBEN
GRISHAM

香江汇时尚餐厅

XIANGJIANGHUI FASHION RESTAURANT

项目名称： 香江汇时尚餐厅
项目设计： 广东星艺装饰集团
项目地址： 广西南宁
项目设计： 李明

本案是一个极具震撼力的现代空间。

乍一看，说不出有何具体的主题，却让人体验到一种全新的奇妙境界：浪漫、温情、简练和时尚，遵循极简主义，表现出新一轮的现代主义与新古典主义。

本案脱离了以往餐厅的风格，这是设计师首要的设计方向。

比如，注重整个空间的灯光，并在天花板上方、四周构造出具有符号韵味的造型效果，又不是简单地以造型取胜，而是以减删的手法，经过构思再提炼，最终以简约的形式把主题暗藏其中，传达出一种幽远的饮食风味。

正如空间色调以稳重的黑为主，少量激情的红为辅，这种黑与红的碰撞对应着冰与火的糅合，同时，又适度地以文化砖、涂料艺术和材质马赛克制作出了抽象的造型或山水般的纹理，这一股股低调的热情、简约的华贵，也荡漾着一脉耐人寻味的海派温情。

如此的组合，交融变幻，身处其中，在享用赏心悦目的夜宴时，便开启了一段宁静的历程。

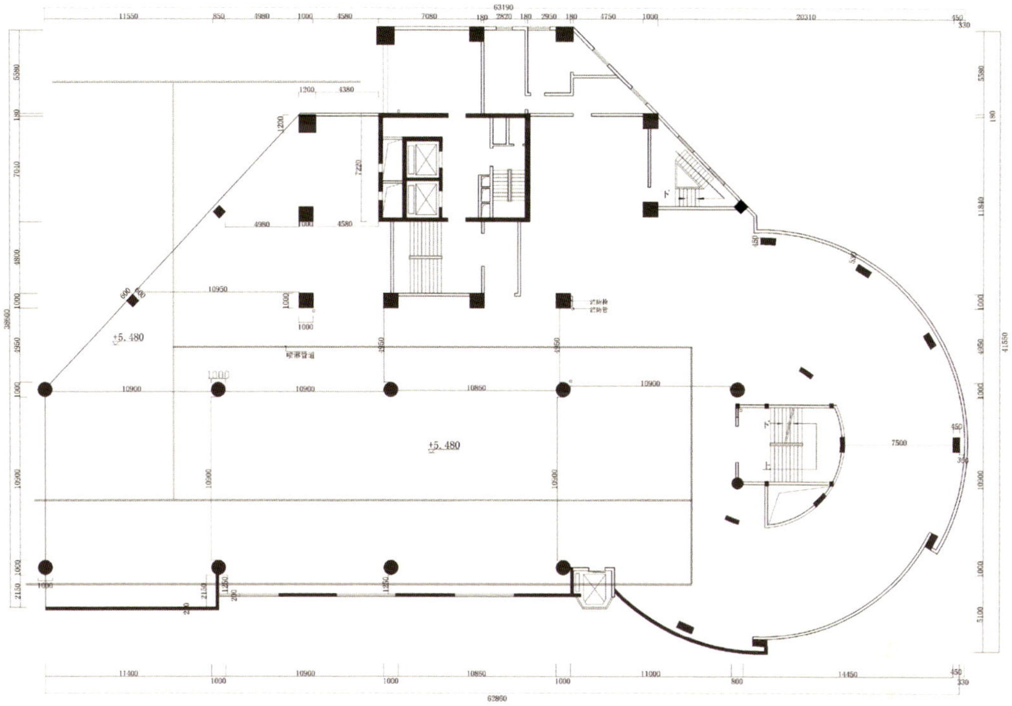

总层高5480mm
风口离地3750mm

原建平面图1：200

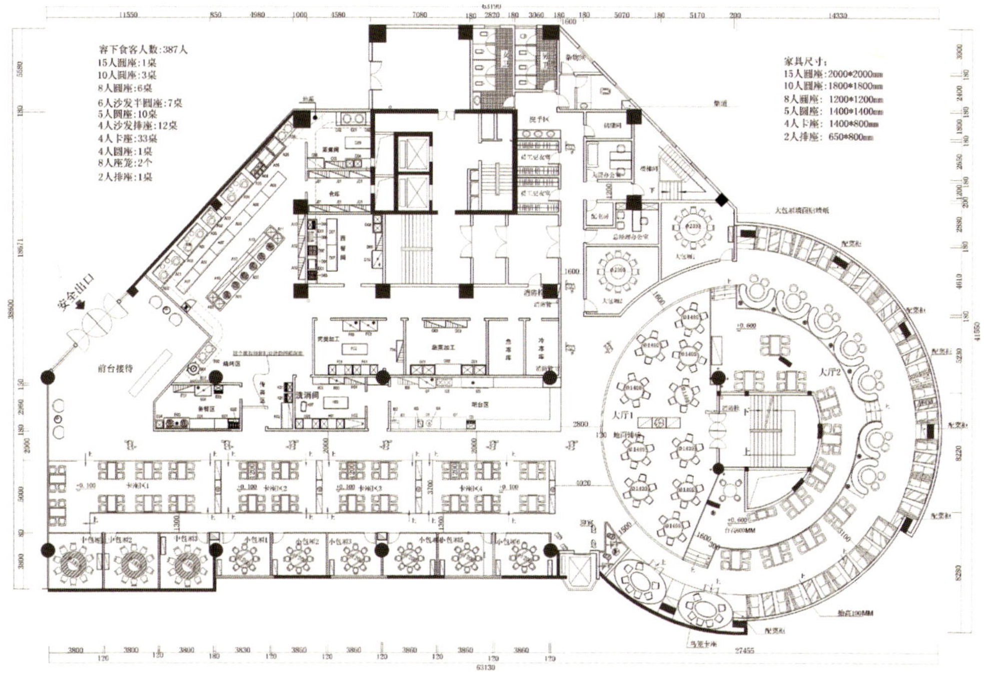

平面布置图1：200

It is a great stunning and modern space.

At the first sight of it, you can't see any specific theme about it but will get a brand-new and marvelous experience in this artistic concept: romance, warmth, simplicity and fashion. For pursuing minimalism, it removes the application of flamboyant materials and complicated craftsmanship, manifesting a new round of modernism and neo-classism.

The restaurant differs from its consistent style, which is the first designing goal that the designer considers about.

For example, more attention is paid to the lights in the whole room and the posture effects of symbols are shown on the ceiling and walls. Instead of simply highlighting the shape to attract visitors, it, by removal and refinement in concepts, hides the theme in the simple design and creates a kind of diverse flavor in diets.

As the room mainly uses black meaning steady, with the supplement of redness of somewhat passion, the combination of black and red accord with the mixing of ice and fire. Meanwhile, on the basis of culture bricks, artistic coating and mosaic of material quality, abstract shapes or texture looking like landscapes are made, where blasts of humble passion and simple luxury spread like ripples of thought-provoking warmth of Shanghai style.

With such a combination varying, one may takes the first step in a course of tranquility, when appreciating the pleasant evening banquet in it.

chapter

② 住宅·工程实景作品

保利凤翔

BAOLIFENGXIANG VILLA

项目名称： 保利凤翔
项目设计： 广东星艺装饰集团
项目地址： 广西南宁
设计师： 覃雷

本案为美式休闲风格，以自然色调为主，白色天花板与蓝色墙漆相衬；别致的小饰品搭配精致的家具，体现了随意轻松的生活方式。它摒弃了烦琐和奢华，强调“回归自然”，使居家空间变得更加轻松、舒适。不论是真皮家具，还是带有岁月沧桑的配饰，都散发着休闲、自然的色彩。

With American leisure style, this scenario focuses on the natural color of white ceilings against blue walls; and the exquisite furniture decorated by delicate accessories present a lifestyle of random and relaxation. It abandons trivial details and luxury and emphasizes the return of nature, making the living space more relaxing and comfortable. Whether the leather furniture or accessories of a long history , all manifests a sense of leisure and nature.

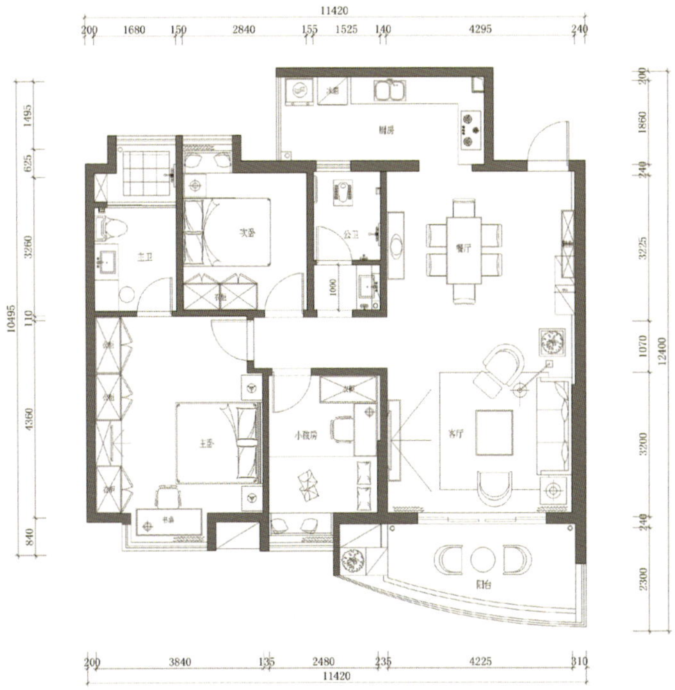

平面功能布置图

SCALE 1:75

大自然花园别墅

NATURE GARDEN VILLA

项目名称：大自然花园别墅
项目设计：广东星艺装饰集团
项目地址：广西南宁
设计师：屈忠涛

本案业主希望将传统欧式风格与个人追求融于一体，打造一种属于自己的古典欧式风格。设计师打破古典欧式风格以金色为主基调的传统，大胆尝试以黑褐色调为主色。黑褐色带来的神秘感，让整个空间呈现出一种低调的奢华，同时也具备贵族气质。另外，业主对于软装的要求很高，从沙发、灯饰、油画、艺术摆设品等都是与设计师共同购置的。

新古典欧式风格典型的设计语言就是注重线条的使用，并强调线条流动变化带来的美感。如在客厅、餐厅中，设计师采用石膏打造吊顶，大量直线条的叠加在视觉上使空间得以拓宽。同时，弧形线条则形成了收敛凝聚的引导效果，让人的视觉更加集中，也让空间在悄然间流露出一种细腻与奢华。过道的多层吊顶配以黑色的镜面玻璃，与墙面的壁纸相呼应，自然地映衬出各自区域的划分。为了凸显客厅挑空区域的大气，设计师采用大面积的木饰面与墙纸互相配合，加上铁艺栏杆的点缀，给人以刚柔相济的冲击力。简约而大气的菱形地面铺贴，则继承了新古典欧式风格的独特别致。二层挑空层的边框，通过大理石罗马柱的承接，强调了墙面与天花色彩的反差，进而突出材质的不同质感，同时也保留原始纹样，很好地体现了古典欧式风格的韵味。黑白色调的反差是这套大宅的亮眼设计，在黑与白的极致运用中，住宅空间的独特魅力显现无遗。

In this scenario, the owner is dying for creating a classical European style of his own, which is integrated by the traditional European style and individual pursuits. Against the classical European tradition to keep gold as its basic color, the designer tries to make dark brown as the main color for the first time. Cause dark brown carries a sense of mystery, filling the room with a humble luxury and enjoying a style of nobility. In addition, with the high demands of householder for soft outfit designs, the householder buys sofa, lighting, oil painting, craft furnishings etc. with instructions of the designer.

The typical European designing language of Neo-Classicism focuses on the application of lines and highlights the beauty of varying lines. For example, when the designer builds sprung ceiling with plasters, the composition of straight lines extends the space in vision. At the same time, curves make effects on convergence and condensation, appealing to visitors more in concentrated visual perception and making the room manifesting a type of exquisite and luxury. The sprung ceilings on the aisles ornamented with black polished glass are consistent with the wall papers, naturally setting off the divided districts. To present the magnificence of spatial living room, the designer uses wood and wall papers together in a large scale, ornamenting these with artistic iron fence, thus forming a blast of impact with hardness and softness. Simple but magnificent paved paste in the shape of diamond succeeds the European uniqueness of neo-classism. In the second floor, the framework, sustained by marble columns, forms a contrast in colors between the walls and the ceiling so as to highlight the diversity of textures in materials. Meanwhile, the primary dermatoglyphic patterns are kept, and thus the classical European style is manifested. The contrast between black and white is the prominent design of this mansion, for the great application of black and white shows all charms of housing.

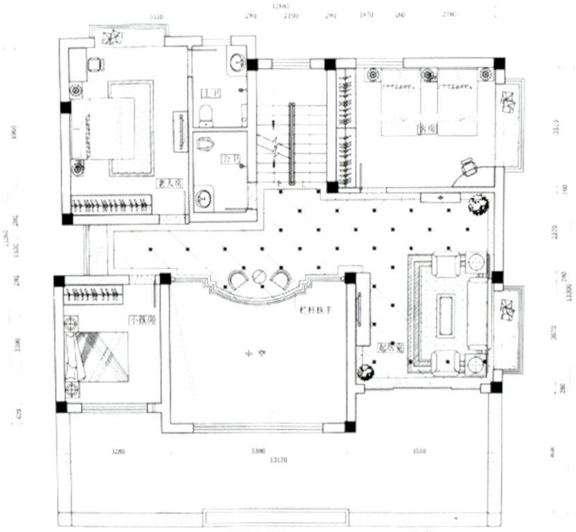

层平面布置图1:75

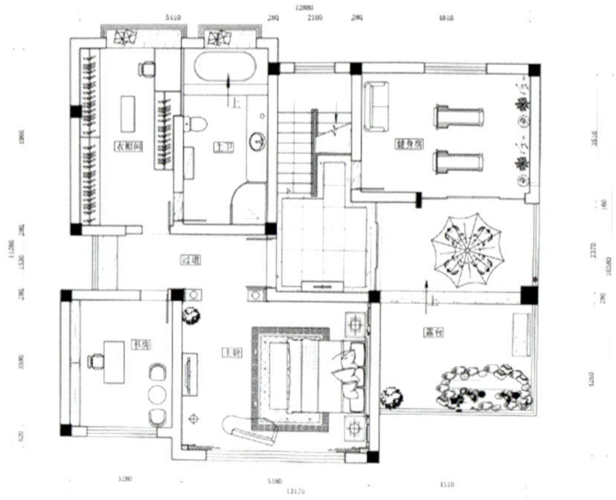

二层平面布置图1:75

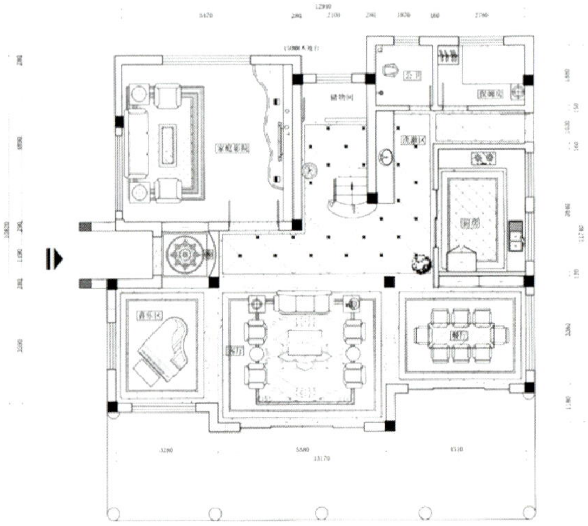

层平面布置图1:75

德利凤岭世家

DELI FENGLING ARISTOCRATIC FAMILY

项目名称：德利凤岭世家
项目设计：广东星艺装饰集团
项目地址：广西南宁
设计师：刘亚　罗芳

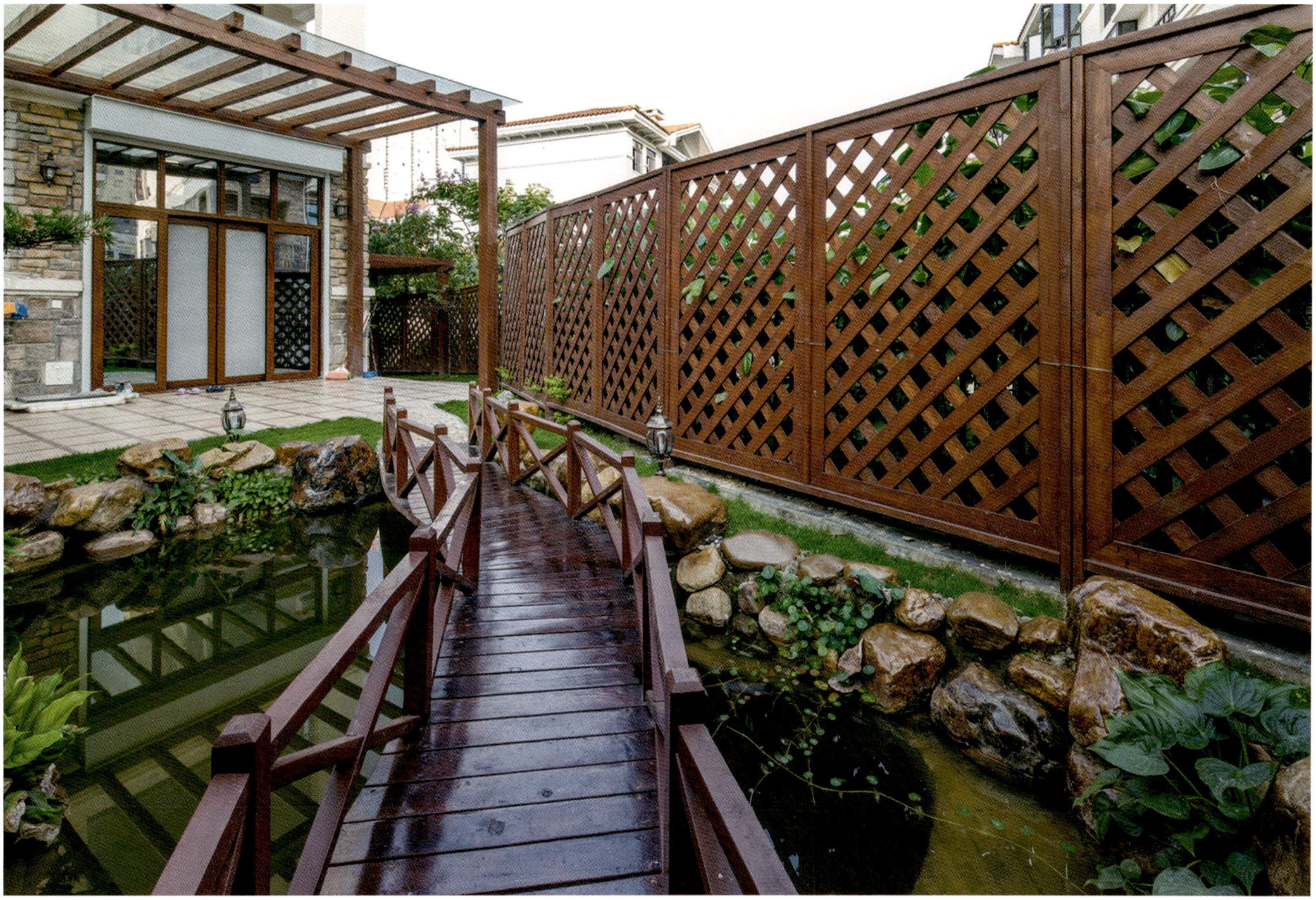

本案地处南宁市东盟商务区内，闹中取静，拥有独特的地理优势。以客厅、餐厅及开放式厨房为中轴线贯穿的设计主次分明，将通风效果利用到了极致，而同步楼梯的修改设计让空间的纵深得到了无限的拉高。白色的欧式家具与色彩艳丽的进口墙布和水刀切割的精致拼花相互搭配，丰富了室内的生活气息，时尚而又华丽。后花园的园林和水系与实木架搭建出一幅美丽的画面，令人心旷神怡。功能齐全的负一楼是健身娱乐的休闲去处，宽大又堂皇的卫浴空间，更是一种身份地位的象征。

Situated in Dongmeng business district of Nanning city, it enjoys unique geological advantages for its tranquility in hustle. With an axis of living room, dining room and open kitchen, it boasts distinction in primary and secondary designs. Meanwhile, the ventilation has achieved best effects while the revisions in the correspondent stairs make the height of the room greatly lengthened. The match of white European furniture, imported flamboyant wallpapers as well as delicate waterjet madellion offer the inside a flavor of life, in fashion and of luxury. The park and water system in the backyard garden as well as wooden shelf consist of a gorgeous painting which is soothing and relaxing. The well-equipped basement boasts a leisure place for exercising and entertainment, where the spacious and stately bathroom becomes a symbol for status.

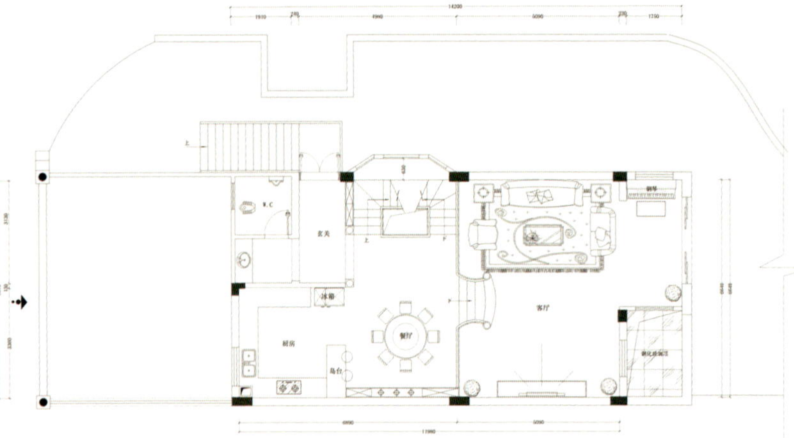

一楼平面布置图1:75

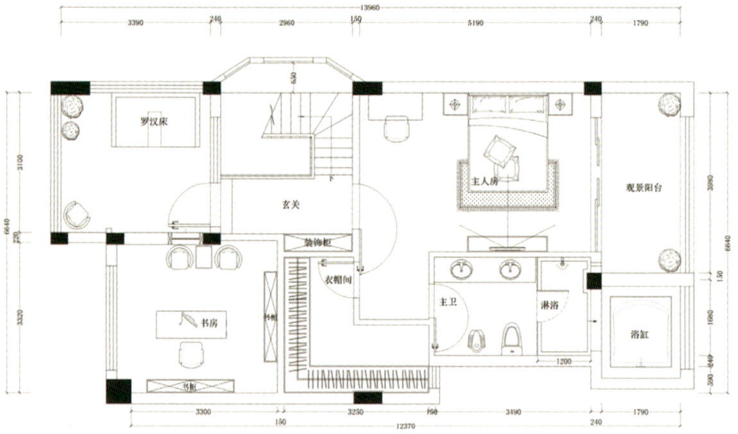

三楼平面布置图1:75

CSK

东岸御花园

EASTERN COAST IMPERIAL GARDEN

项目名称：东岸御花园
项目设计：广东星艺装饰集团
项目地址：广西柳州
设计师：杨复民　张如意

本案的设计做了很大的平面改动，在整体空间的表达上，将欧式新古典风格与美式乡村风格相融合，自然温馨却不失奢华之气。空间的层次性，从玄关到餐厅的过渡，通过门廊、木色、地面的材质改变，让空间有一定的递进性和层次感，并让不同空间之间有了一定的联系。客厅和餐厅之间也是通过视觉感官和地面材质的改变让彼此相呼应。地面采用木地板，让客厅的空间更加温馨，有家的感觉。房间采用简单的方式处理，使空间留白，在未来可以添置一些客户真正需要的东西，温馨舒适而不繁杂浮躁。

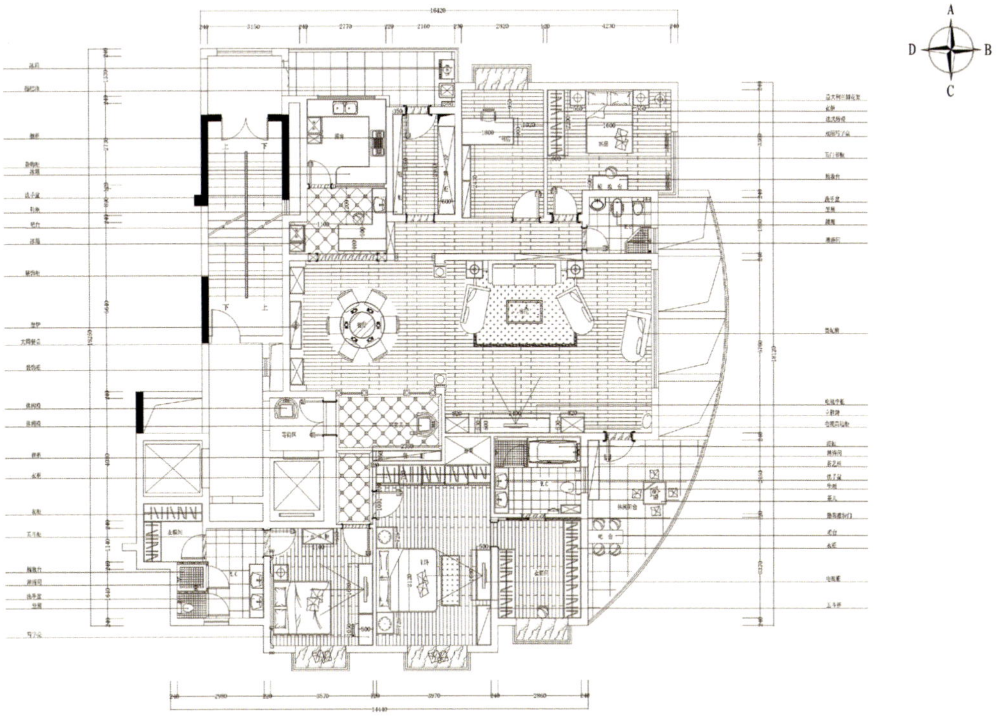

平面功能布置图1:75

With general layout changed in a large scale, the design, in the manifestation of space, tries to combine European Neo-Classicism and American country style and thus to pursue nature and warmth as well as luxury. The spatial hierarchy, via the transition from the porch to the dining room, makes it multistep and layer by layer through changing the materials of the porch, wood color and the floor, which has relations in diverse spaces. And the living room and dining room go together for varied visual sense and floor materials. The floor made of wooden board; the living room presents much warmer, making one feel like at home. Besides, with decorations in a simple way for making it spacious, there is much room for guests to place those indeed valuable goods, in order to create a warm, cozy but less miscellaneous and impulsive atmosphere.

东盟中央城

DONGMENG CENTRAL CITY

项目名称：东盟中央城
项目设计：广东星艺装饰集团
项目地址：广西南宁
设计师：凌立成

本案为一个常规的两厅四室的户型，在接到这个户型之前，业主提出美式风格的设计要求，但是这个户型结构上相对偏小，而美式家具尺度偏大，这对设计也算是一个考验了：如何在一个小空间中摆入“大块头”？本案所呈现出来的就是经过设计师改造后变得更加宽敞、光线更加充足的空间设计。首先，在结构上通过增补面积做了些小处理，构造出了一个巧妙的入户玄关，将原有的厨房和餐厅合并为一个空间来使用，把原有的生活阳台改造为中厨房，在餐厅增加一组收纳橱柜作为备用的西餐柜，以弥补中厨的不足。在色彩以及风格设计方面，融合了业主提出的西方美学思想，并结合其生活理念进行设计，以温馨的黄色和赏心悦目的蓝色为主调，为整个空间增添了丰富的视觉效果。

It is a place in regular layouts of two dining rooms and four bedrooms while, before taking over the project, the householder put forward the demands for designs in American style. But, with the contrast between this comparatively small layout and large American style house, there is no doubt that it is a test in designs: How to place a large hunk in the little room? Here, you may see the more spacious design alive with sunlight after revisions of the designer. First of all, sort of treatments are made to increase the area in structure, thus an ingenious porch for entrance is built. And to compensate for the lack of Chinese kitchen, the designer integrates the original kitchen and dining room, transforms the original balcony into Chinese kitchen and places another storage cabinet as a spare western food cabinet in the dining room. Apart from this, in the designs of colors and styles, combining the householder's advice of western aesthetics, making designs according to his life concepts, the designer gives priority to the warm yellow and pleasant blue, thus bountiful visual effects of the whole room are strengthened.

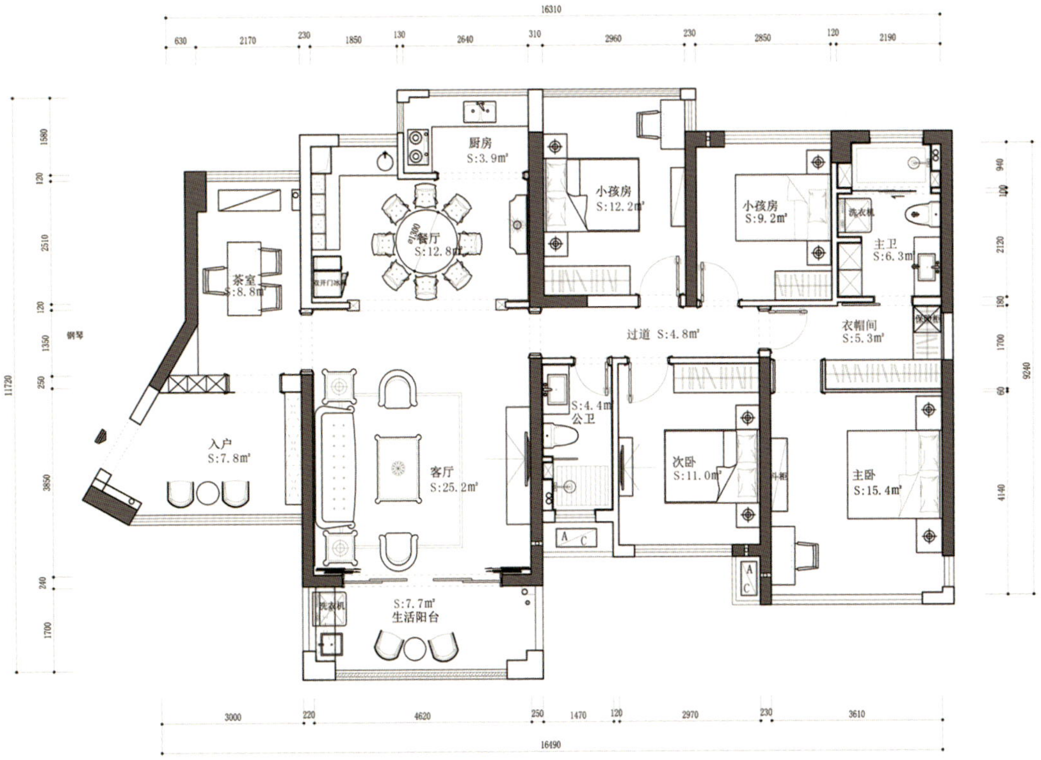
16310
厨房
S:3.9㎡
小孩房
S:12.2㎡
小孩房
S:9.2㎡
主卫
S:6.3㎡
洗衣机
餐厅
S:12.8㎡
茶室
S:8.8㎡
过道 S:4.8㎡
衣帽间
S:5.3㎡
入户
S:7.8㎡
客厅
S:25.2㎡
S:4.4㎡
公卫
次卧
S:11.0㎡
主卧
S:15.4㎡
S:7.7㎡
生活阳台
洗衣机
16490

凤景湾

FENGJING BAY

项目名称：凤景湾
项目设计：广东星艺装饰集团
项目地址：广西南宁
设计师：杨云星　陈汐

家居设计有三种境界：一是设计一套房子；二是设计一个家；三是设计一种生活方式。不同的心境会收获不同的结果，而第三种境界就是我们一直追求的。我们经常在思考和想象一家人休假的最佳状态：一个大大的书房，可以在里面随意地翻阅书籍。一个富有童趣的儿童房，可以让小孩尽情玩耍。一个宽阔的主卧，主人可以在卧室里懒散地走动。一个大大的阳台，让人独享时光，可以在那安静地画画，安静地品茶……这样的设计思想在本案中多多少少能得到一些实现。

There are three realms of home design, which are respectively designing a house, a home and a way of life. In different moods, different design effects will be embodied, and what we have always pursuing is the third realm-to design a way of life for the owner. In the design, it integrates the reflection and imagination on lifestyle: people can read books at their options; kids can play to their hearts' content in a childish children's room; the owners can happily walk in the wide master bedroom; people also can quietly sip the tea and enjoy the joy of solitude.Such kind of design is somewhat reflected in this scenario.

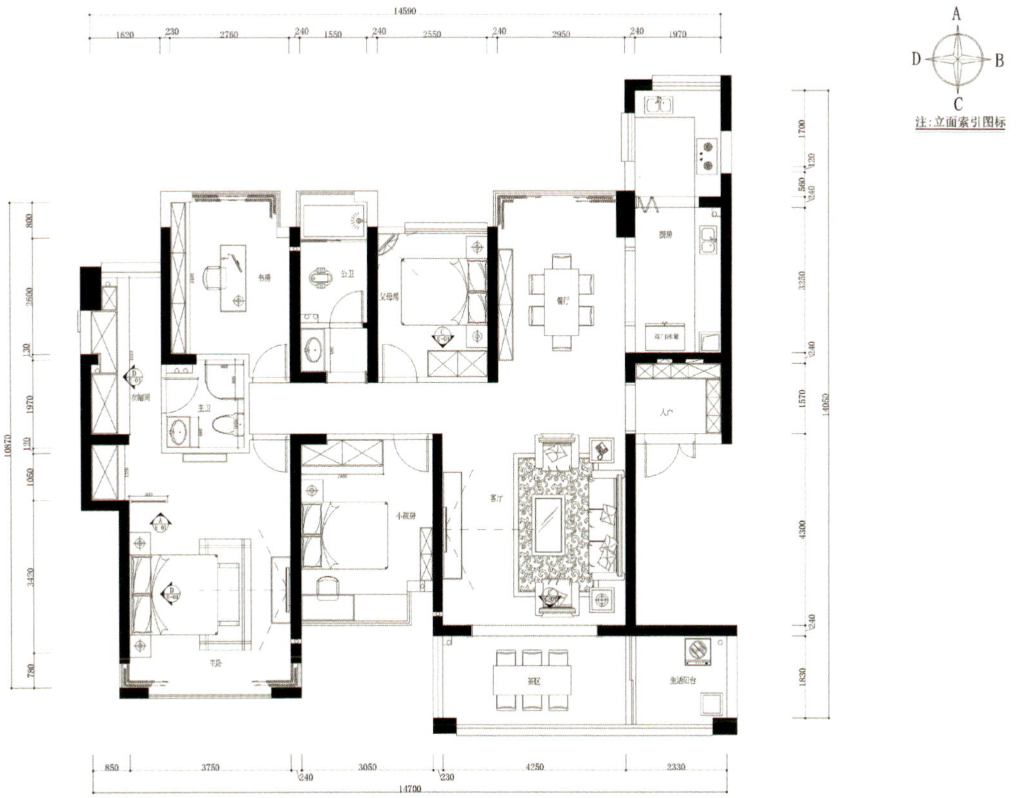
14590
1620
230
2760
240
1550
240
2550
240
2950
240
1970
书房
公卫
父母房
餐厅
厨房
入户
衣帽间
主卫
小孩房
客厅
茶区
生活阳台
850
3750
240
3050
230
4250
2330
14700
A
B
C
D
注:立面索引图标

凤岭一号

FENGLING NO. 1 RESIDENCE

项目名称：凤岭一号
项目设计：广东星艺装饰集团
项目地址：广西南宁
设计师：覃雷

D&D
Furniture
LUXURY HOTELS
The Hotel Book
Arletta
SPACE

D&D
Furniture
LUXURY HOTELS
The Hotel Book
Arletta
SPACE

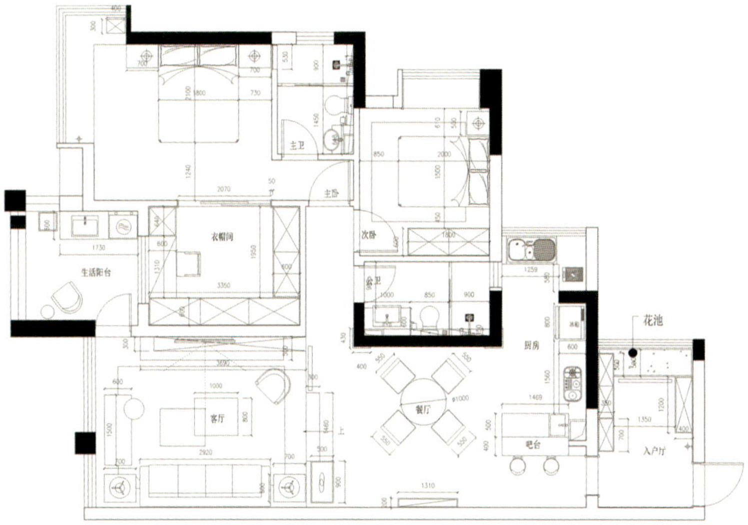

业主何先生是年轻的单身精英人士，追求简单舒适的生活方式。整个空间设计没有过多的修饰，干净利索的处理，时尚的黑白灰搭配，亮色的软装点缀，呈现了一个现代时尚的单身男士住宅。

The owner, Mr. He, of this building is a young single elite and seeks for a simple comfortable lifestyle. Without much decoration, the whole space is decorated with black and white materials. With some bright-colored soft embellishment, the residence takes on an air of modern fashion of a single man.

海华 · 东盟公馆

HAIHUA · DONGMENG RESIDENCE

项目名称：海华 • 东盟公馆
项目设计：广东星艺装饰集团
项目地址：广西南宁
设计师：杨艳

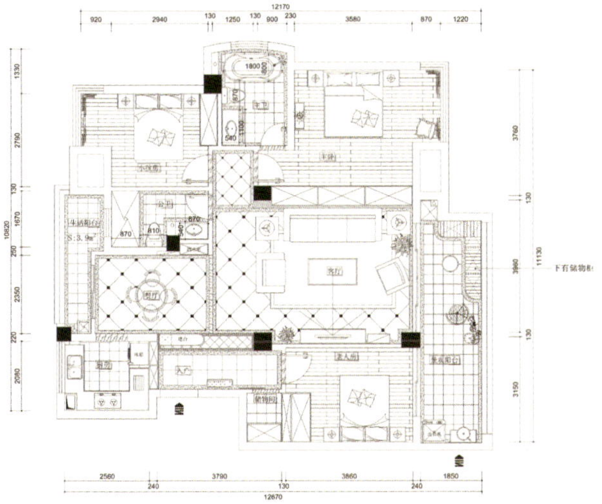

本案是“80 后”妈妈 130 平方米的美式温馨家，业主梦想中的生活是这样的：坐在独立的吧台上，在一个宁静的午后，自己细细磨上一杯咖啡，听着悠扬的音乐，看着孩子在沙发上懒懒睡去。偶尔兴致勃发还能捧起一本时尚杂志，了解最新流行趋势。这就是“80 后”妈妈渴望在快节奏工作之余拥有的一个属于自己的休息空间。

This scenario is a 130m² American style house of a "post 80s" mother. The owner’s dream is like this: on a quiet afternoon, she sits on an independent bar, with a cup of finely brewed coffee, listens to the melodious music and sees kids lazily sleep on the sofa. Sometimes, she reads a fashion magazine to know the newest fashion trend. The “post 80s” mother longs for owning a leisure space which belongs to herself after fast-pace work.

THE CLASSIC

翰林雅筑

HANLIN BUILDING

项目名称： 翰林雅筑
项目设计： 广东星艺装饰集团
项目地址： 广西南宁
设计师： 周兵　兰瑞严

本案为一家三口起居空间，因业主喜美式家具，房屋公共空间的面积又不是很大，最终将风格定位为欧式与美式交融合一，创造出一种戏剧性的时尚美学观，营造出自然而开阔的美式空间氛围，以及彰显尊贵气派而不张扬的浪漫欧式情调。从空间整体的布局到细节之处的精心点缀，都体现了设计师人居环境与自然和谐相处的设计理念。整体空间布局浑然天成，华美且实用。在材料的使用上，从晶莹剔透的米黄大理石、咖啡网纹大理石、硬包，到天花板的欧式线条，以及墙纸和水晶吊灯，到处都弥漫着一种奢华与浪漫的气氛。别致的家具是身份的最好象征，用现代的手法去重新演绎经过几百年文明沉淀的美式家具，一点一线间都透露出优雅的气质。

This scenario is designed for a three-member family. The design style is determined as the communion of European and American styles, because of its householder's preference for American furniture and somewhat the smaller area for public activities. Besides, it adds a kind of dramatic and fashionable aesthetics, which creates an American spatial atmosphere with a sense of natural and spacious and the European romance of nobility and humbleness. From the spatial layouts to the meticulous decorations in details, the designer's concept of harmonious life between living environments and nature is presented. As to the spatial layout, it looks like being made by the nature, exquisite and practical. As for the use of materials, an atmosphere of luxury and romance emanates hither and thither, from the glittering and translucent beige marbles, emperador marble patterns, hard rolls, to European lines on the ceiling, wallpapers and chandeliers. The exquisite furniture is the best symbol of status, as is shown in American furniture which applies modern methods into deducting hundreds of years' civilization and of which elegance presents in a dot and line.

主卫
主卧
小孩房
次卧
客卫
餐厅
客厅
阳台
厨房
入户

华商国际欧洲城 A 样板房

HUASHANG INTERNATIONAL EUROPEAN CITY MODEL ROOM A

项目名称： 华商国际欧洲城 A 样板房
项目设计： 广东星艺装饰集团
项目地址： 广西玉林
设计师： 许舰 黄国展

平面布置图 1:75

本案华丽流畅的线条，配合富有古典风情的色彩，完美展现优雅、时尚的不凡气质，让生活形态和美学意识转化为一种高雅的品位。

This scenario uses smooth lines with rich classical colors to create an extraordinary temperament of elegance and fashion, which has perfectly transformed lifestyle and aesthetics consciousness into a kind of refined taste.

华商国际欧洲城 B 样板房

HUASHANG INTERNATIONAL EUROPEAN CITY MODEL ROOM B

项目名称：华商国际欧洲城 B 样板房
项目设计：广东星艺装饰集团
项目地址：广西玉林
设计师：许舰　盛强

现代简约风格，巧妙的留白，看似简单，背后却凝聚着设计师力求美观而实用的独特匠心。

The modern concise design with the artful blank-leaving of the space, looks simple, while behind it, the unique originality of aesthetic and practicality is converged by the designer.

平面布置图 1:75

REMY MARTIN
CLUB

PaperArt
PaperArt

HEDGE HOGGING
Arletta

BOB MARLEY
SPANISH FOOTBALL CLUB
HEDGE HOGGING
Arletta
Arsenal
Aesop's Fables

商务礼仪
Manchester United Football Club

汇东逸墅

HUIDONG LEISURE HOUSE

项目名称： 汇东逸墅
项目设计： 广东星艺装饰集团
项目地址： 广西南宁
设计师： 范建国　黄林妮

7 号院位于南宁市东盟商务区核心地带，闹中取静，整个别墅区建筑规划呈现出浓郁的现代东方气息。在室内设计的风格上，经与业主充分沟通，设计师将其定位为现代中式风格。在功能使用上，负一层为娱乐室、收藏室、红酒窖和私人 SPA 室，满足业主与亲朋好友的交流娱乐需要；一层为客厅、饭厅、厨房等主要起居公共空间；二层、三层为卧室区，并配备专用的茶室和健身房。现代中式风格并非完全意义上的复古明清，而是通过重新演绎中式风格的特征，表达对清雅含蓄、端庄风华的东方式精神境界的追求，因此在设计上不仅包含了中式的传统特点，还融入了一些现代色彩和灯光处理手法。元素简化后的木质顶棚造型，既突出了收藏室和品茶区的文化属性，又很好地契合了现代的工艺处理。灰蓝色的现代绒面窗帘柔化了红木家具的深厚，又在色彩上起到了冷暖对比的作用，配合 LED 的隐藏式灯光，使整个空间温暖柔和，打破了传统中式风格沉重压抑的氛围。通过对传统中式木梁的分解重组，设计了独特的天花板斜线木质收口系统，既融合了如中央空调管线及 LED 隐藏灯光的布局需要，又突出了简练的中式线条造型。而无处不在的名人字画则作为空间点缀，使整个别墅透露着淡雅的文化气息。

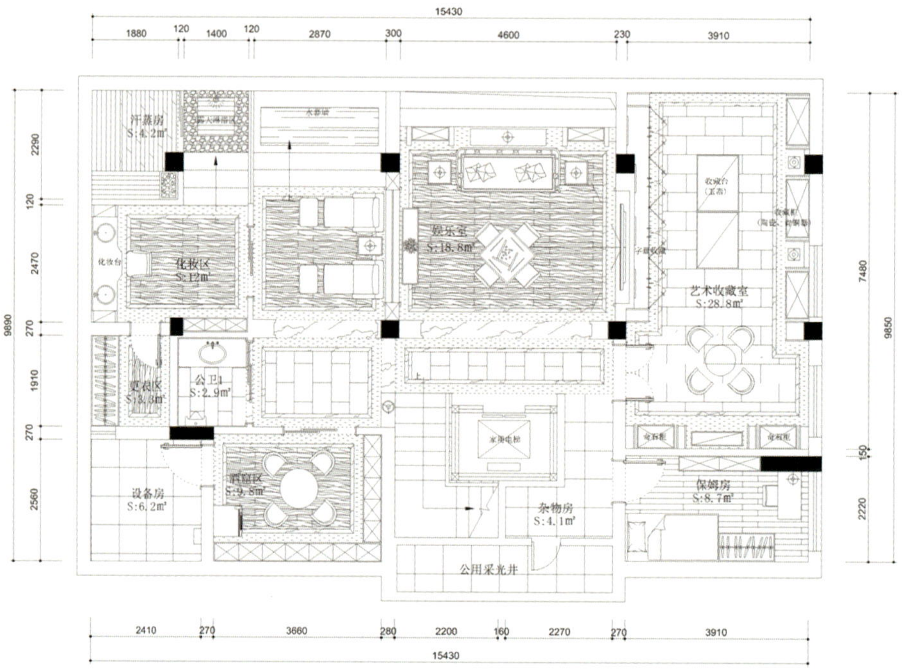
15430
1880
120
1400
120
2870
300
4600
230
3910
汗蒸房
S:4.2m²
水景墙
娱乐室
S:18.8m²
化妆台
化妆区
S:12m²
艺术收藏室
S:28.8m²
更衣区
公卫1
S:2.9m²
设备房
S:6.2m²
酒窖区
S:9.8m²
杂物房
S:4.1m²
保姆房
S:8.7m²
公用采光井
2290
120
2470
270
1910
270
2560
9890
7480
150
2220
9850
2410
270
3660
280
2200
160
2270
270
3910
15430

House 7 is located in the core area of Dongmeng Central Business District of Nanning City, a place of tranquility in the hustling city, and architectural plans of the whole villa district present a strong scent of modern orient. As for the style of interior design, via communications with the householder, the designer designates the style as modern Chinese style. As for its function, the basement works as recreation room, boxroom, wine cellar and personal SPA room for the householder to entertain with the relatives and friends; the first floor as public area mainly for living such as living room, dining room, kitchen and so on; the second and third floor as bedrooms area, well-equipped with specialized tea houses and the gyms. Rather than the complete quaint style in Ming and Qing dynasty, modern Chinese style, by deducting the Chinese characteristics again, expresses the pursuit in oriental spiritual realm of light elegance, implicitness and grace, owing to which the design includes not only traditional Chinese features but some modern colors and approaches to dispose lighting. The shape of wooden roof after the simplicity of elements not only highlights the cultural feature of boxroom and tea house, but also satisfies the modern crafts treatment. Gray-blue modern suede curtain makes the hard rosewood furniture soft, forming a contrast between coldness and warmth in colors. Besides, with the LED hidden lighting, the entire house presents warmth and softness, breaking the depression of traditional Chinese style. Through decomposition and restricting of traditional Chinese wooden beams, the designer creates a unique ceiling wooden closing system, which not only meets with the layout requirements in pipelines of central air-conditioners and LED hidden lighting but emphasizes the concise shapes of Chinese lines. However, the calligraphy and paintings hither and thither decorate the house, making the entire villa alive with slightly cultural atmosphere.

南湖名都

NANHU TOWN CENTRE

项目名称：南湖名都
项目设计：广东星艺装饰集团
项目地址：广西南宁
设计师：谢树声

老子说：“大音希声，大象无形”，如履如境，简而有韵，精而有致。所有一切的视觉形态在这里似乎都变得如水般清澈起来，透出那一份隐隐的中式情怀。没有高贵的石材，没有华丽的装饰手法，唯有那质朴地板幽幽地透着一股木香，让人不知不觉淡定神宁。形态优美的桌椅，做工精良的装饰线条，端庄而稳健，无不在传递着主人修身养性的生活态度。一个家承载更多的不是富丽堂皇的装饰，而是一份别致有韵、温馨可依的居所。

As the famous saying in Lao-Tze: "Great sound is hard to hear, large form is beyond shape." When you step into the house, you would feel like wandering in the paradise, which is simple but charming, refined but elegant. All the things at sighet are like the clean water, which creates a sense of Chinese feelings. This scenario has no expensive stone, nor luxuriant decoration method, but uses the plain floor, beautiful tables and chairs and exquisite decoration lines to create a life atmosphere making people calm and chillax. A home bears not only gorgeous decoration but a chic and warm residence.

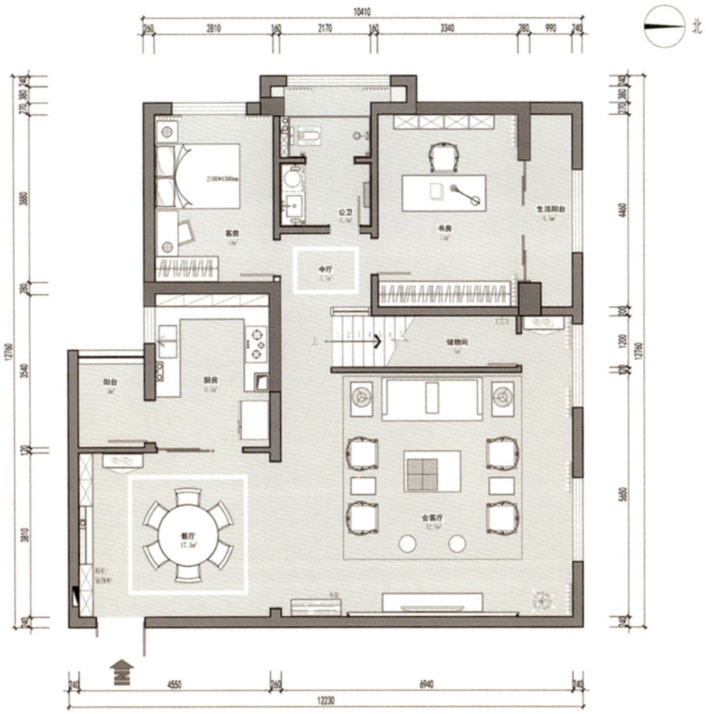

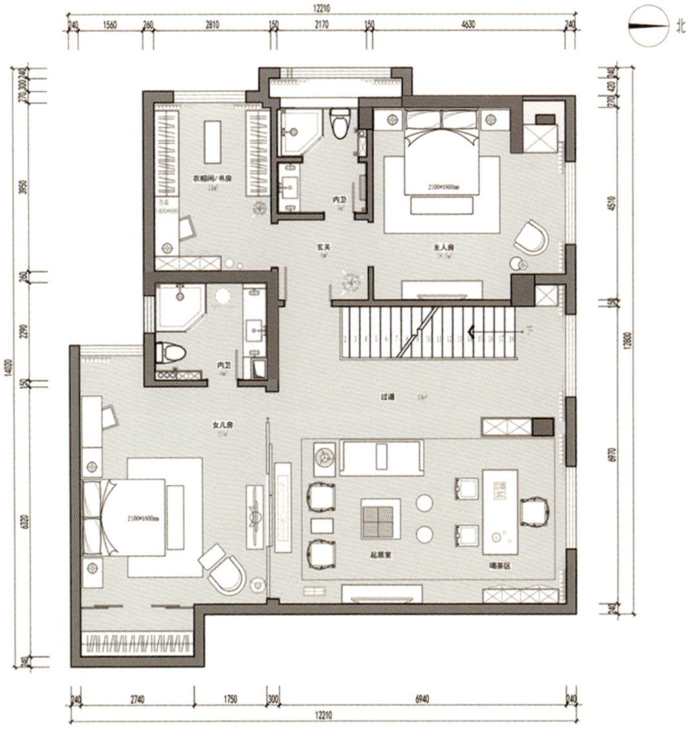

南湖名都 2201

NANHU TOWN CENTRE 2201

项目名称：南湖名都 2201
项目设计：广东星艺装饰集团
项目地址：广西南宁
设计师：屈忠涛　钟宁章

“南湖名都”处于南宁环南湖区域规划的综合居住片区，项目周边已建成南湖广场、金湖广场、国际会展中心等多个市政项目，区域优势明显，不仅拥有极佳的地段，更是好地段中极为稀缺的地块，建筑外观为经典的灰白配色。

Situated in the complex of residential areas planned by Nanhu district of Nanning City, with multi-municipal projects such as Nanhu Plaza, Jinhu Plaza, International Convention Center and so on. "Nanhu Town Centre" enjoys distinctive geological advantages, accounting for the remarkably wonderful lands which are extremely rare amid the lands and making classic grey and white as its color of the appearance of architectural buildings.

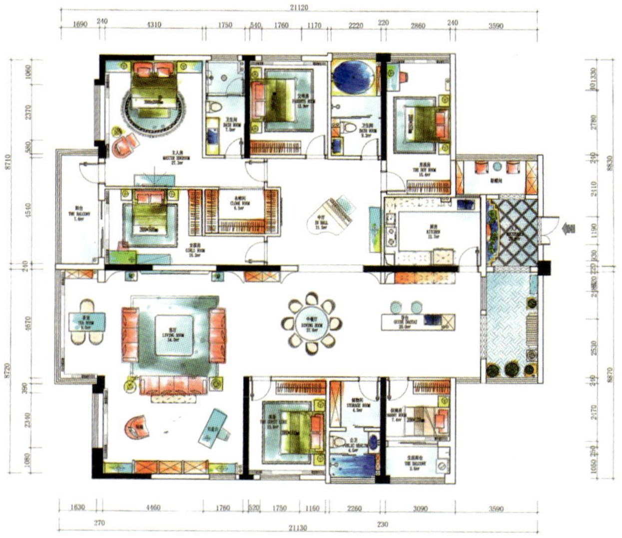

平面布置图 1:100

本案试图寻求一种建筑与室内融合共处的思想理念。西方文化的审美表象和传统深处不可磨灭的意识相交杂，也是部分业主心理的真实反映。对空间进行立面处理，强调对称、平衡和穿透，并最大限度地引入室外的自然园林，让其与入户门厅的音韵设计前后连贯。宽敞的室内为我们展示的是西方文化的景象。各种简意的雕塑和挂画，既显示了主人的收藏品位，又在室内陈设中形成多重构图，让人从每一个角度望去都赏心悦目。每一个空间，在西方文化的格调上都展现得恰到好处。客厅背景更是别有一番韵味，海纳百川与窗外的南湖景相呼应，和简洁的沙发床艺一同谱写出悠然雅致的欧式风情。

This scenario tries to seek for a concept of integrating buildings to their interior layouts. The aesthetics of western culture interchanges with everlasting memories in Chinese tradition, also reflecting on the psychologies of some householders. As for the spatial facade design, the symmetry, balance and penetration of space are emphasized. The natural park outdoor is connected with the room in full measure, which maintains consistency with the lingering charm of the entrance hall. On the one hand, the spacious interior of building presents a picture of western culture; on the other hand, a variety of simple statues and hanging pictures manifest the householder's taste in collecting and consist of multiple compositions on furnishings, thus forming a pleasant sight when seen from any perspective. With characteristics of western culture, every single room presents a good looking. The elegant and exquisite setting of the living room, of which the blue sea is consistent with the Nanhu scenery outside the window, shows European style of leisure and elegance with the concise sofa and beddings.

荣和大地 17A

RONGHE BUILDING 17A

项目名称： 荣和大地 17A
项目设计： 广东星艺装饰集团
项目地址： 广西南宁
设计师： 徐橘　陈思成

本案为美式风格。美式风格以宽大、舒适，杂糅各种风格而著称，是崇尚自由的代表形式。本设计承袭美式建筑艺术和美学神韵，内饰回归美式最正统的浪漫典雅，且对传统重新做了尝试与思考，化繁为简，吐故纳新，将美式风格的端庄与简约，与当下中国审美情趣合二为一，实现端庄与现代、传统与创新的和谐之美。

This scenario is in American style renowned for magnificence and coziness, sucking the marrow from all kinds of styles, which is a type of pursuing freedom. It not only takes over the art of American architecture and aesthetics, with the pure American romance and elegance in decorations, but makes new tries and considerations on the tradition by simplifying the flamboyance, getting rid of the old-fashioned and taking in the innovations. During the process, combination between American style of elegance and simplicity and current Chinese aesthetics presents the beauty of harmony of integrating elegance and modernity, tradition and innovation.

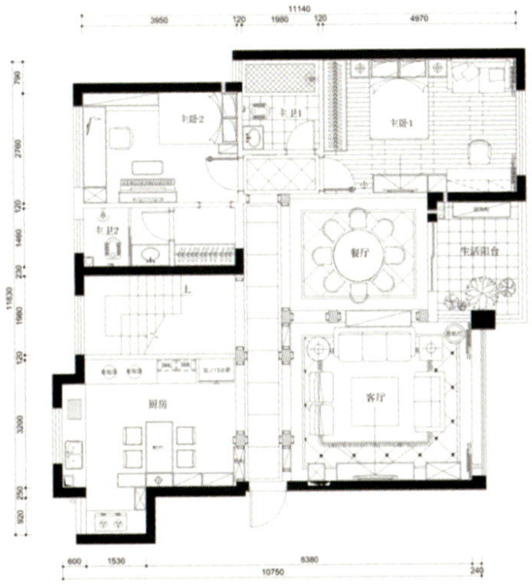

一层平面布置图 1:75

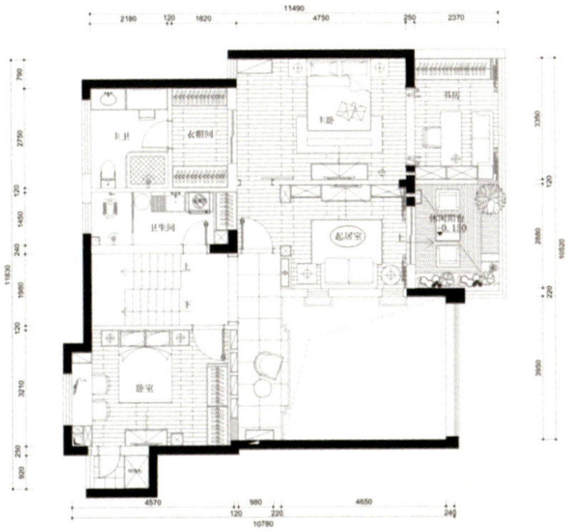

二层平面布置图 1:75

整个空间陈设注重气质，留意精神世界。室内一共三层，格局以中轴线对称布局。客厅作为入户第一空间，由造型优美的美式廊道导引，空间主体色调是气质不凡的麦芽白，颜色耐看而不做作。完美的廊道拱形、特别订制的复古美式吊灯、极具现代感的优雅家具陈设，以及气质典雅的绒布窗帘，于空间中产生了和谐之美，气质凝练而彰显气度。在客厅占有一席之地的美式壁炉电视背景墙彰显了主人的品位和对悠闲自得的生活的美好追求。墙体的材质均采用了弹涂硅藻泥和硅藻米洞石，质感柔和、温润、均匀。材质的选用体现了主人在细节上的不断追求。

The house highlights the atmosphere created by the decorations and the connections with spiritual life. With three floors, the house layouts are symmetric along the axis. As the first place for entrance, the living room lies an American gallery of gorgeous shape, with malt white of brightness as the main color, a good-looking and natural color. Here a beauty of harmony grows amid perfect gallery arch, specially customized classical American chandeliers, and elegant furnishings alive with modernity and elegant velvet curtains, presenting a style of conciseness and tolerance. Besides, in the living room, the setting wall of American TV in fireplace tells the householder's taste and the pursuit of a leisure life. Based on diatom mud mudskipper and diatom mud travertine meters, the materials boast the qualities of softness, smoothness and equation, towards which the attitude manifests the householder's pursuit in details.

餐厅对外的休闲阳台中的墨绿色装饰柜在空间中产生了冲突之美，大胆的用色和美式传统造型互相碰撞产生了别样的视觉享受。屋顶花园的设计给繁忙的都市生活增添了一个悠然自得的空间，浓郁的绿植，美式风格的露台，将室内与室外共同点缀在自然中，形成很大的活动空间。空中菜园的设计给整个空间增加了更多休闲生活的趣味性，早春黄瓜西葫芦，盛夏辣椒甘蓝菜，延秋莴笋大白菜，越冬萝卜和生菜，每个季节都是一种享受，一种多么悠闲的美式生活情趣。

On the balcony for leisure out of the restaurant, there stands a dark green cabinet for ornament, making a contrastive beauty against the house, with the collision between bold color and traditional American shapes generating a brand-new visual effect. The design of the roof garden offers a leisure place in the hustling urban life, forming a large-scale activity space with exuberant green plants and American-style balcony dotted in the exterior and interior of the house. With this design, the space is becoming more interesting for leisure, with cucumbers and cucurbita pepos in spring, peppers and cabbages in summer, asparagus lettuces and Chinese cabbages in autumn as well as turnips and lettuces in winter, one may feel delighted and joyful in every single season for such a leisure American life.

美式往往不追求简单的协调，而是崇尚冲突之美。在设计上讲求心灵的自然回归感，给人一种扑面而来的浓郁气息。开放式的空间结构、随处可见的花卉和绿色植物、造型简约的家具……所有的一切从整体上传达出一种悠然自得的生活态度。

Rather than seeking for the simple coordination, American style prefers the beauty of contrasts. In designs, the belongings to nature in the depth of heart are advocated in order to create a strong atmosphere. With open spatial structure, flowers and green plants hither and thither, and furniture of concise shapes… a leisure attitude towards life is expressed in all of these.

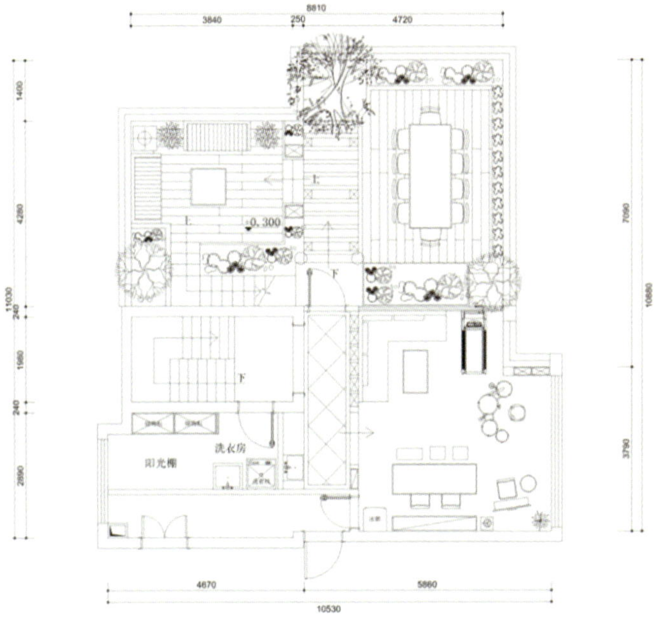
8810
3840
250
4720
1400
4280
±0.300
11030
240
1980
240
2890
7090
10880
3790
洗衣房
阳光棚
4670
5860
10530

山水映像

LANDSCAPE IMAGE

项目名称：山水映像
项目设计：广东星艺装饰集团
项目地址：广西河池
设计师：范建国　许舰

作为一栋拥有五层垂直空间，室内面积达到 1680 平方米的别墅，我们已经不能简单地把它定义成传统的住宅空间了。在与业主充分沟通之后，设计师决定用精品酒店的设计思路来规划别墅的室内空间。通过展示空间、休闲娱乐空间、卧室休息空间、日常起居空间这四大功能空间将整体空间有序地分割并联系在一起。现代东方的简练设计贯穿始终，通过天然石材和现代的灯光处理手法来展现河池地区独有的山水自然风貌。五层中空的楼梯厅空间感十足，铜制的条栅隔断如瀑布飞流直下，甚为壮观。一层厅底的玉石球造景稳定了整个中空空间的基础。线条的各种组合以材质和光影的形式出现在各个功能空间。凹凸有致的格栅书架、优雅整齐的竖向条栅都很好地贯彻了线性在空间中的序列。在色彩的搭配上，除了一层展示空间采用了沉稳的咖啡木色调外，其余空间均以明亮清爽的自然浅木色为主，提升了空间的居住舒适感。空间中的各类木制作及金属品均实行工程定制、现场安装的方式，提高了整体的品质感和细节呈现度。

As a five-storied vertical villa with the interior area of 1,680 square meters, we couldn't simply define it as the traditional residential space. After fully communicating with the owners, designers decide to use the design idea of boutique hotel to plan the villa's interior space. They separate the whole space orderly and make it connective by four functional spaces, namely, exhibition space, leisurely space, bedroom space and daily routine space. The modern oriental simple design runs through it. It shows the unique scenic natural view in Hechi Region by natural stone and modern lighting dealing ways. Its five-storied hollow stair hall has sheer sense of space, coppery grid partitions are like waterfalls flying straight down, which is very spectacular. The jade ball landscaping with one-storied floor stabilize the base of the whole hollow space. The various combination of lines appear in all functional spaces in the form of texture and shadow. Grill bookshelves and vertical bar screens well carry out linear sequence in space. On the match of color, besides the one-storied exhibition space uses the color of coffee wood, other spaces give priority to bright and fresh natural light wood, which promotes the comfortable living sensation in spaces. All kinds of wood manufacture and hard goods implement the way of project customization and field installation, which enhances the whole sense of quality and presented degree of detail.

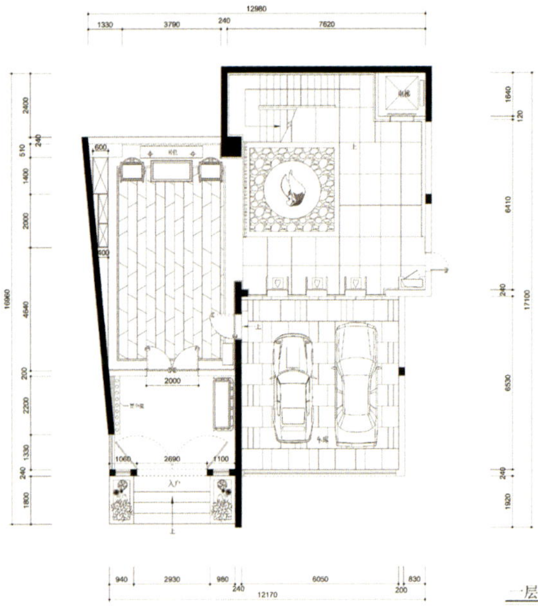

一层平面布置图 1:100

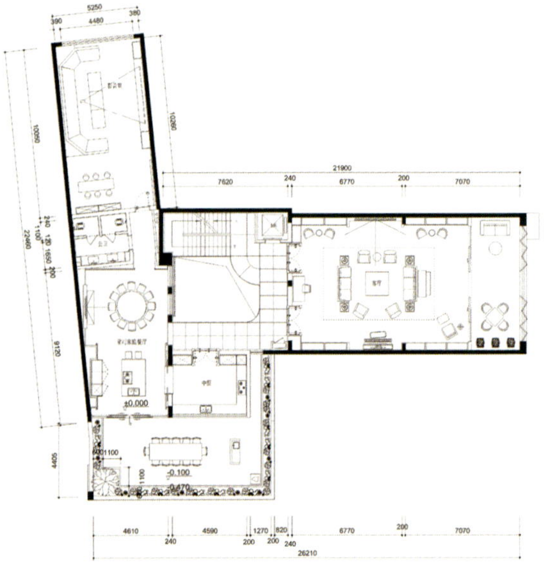

二层平面布置图 1:150

山与城 332 号公寓

MOUNTAIN AND CITY NO.332 APARTMENT

项目名称：山与城 332 号公寓
项目设计：广东星艺装饰集团
项目地址：广西桂林
设计师：张抗

如果说家具的款式能够体现空间的气质，那么色彩的搭配就能丰富空间的表情，而灯光则能活跃空间的气质，本案的设计正好印证了这几点。本案为 99 平方米小三房，因此在平面设计上，始终追求空间和功能的实用与完美，大胆改变原建筑的平面设计，充分利用建筑空间可延展或可利用的位置进行合理的设计再利用。其中，由客厅阳台改造设计后的空间涵盖了收纳区、休闲区及工作区三种功能；设计造型上以现代简约的设计手法融合复古风格的家居及软装饰；色彩搭配大胆运用色块的强烈对比丰富空间的层次感，让空间在复古的优雅中不失青春向上的活力。同时，灯光的设计更是在功能实用至上的基础上，让空间的明度及纵深更加丰满。小空间大利用，设计对于生活不仅是一种装饰，更是一种需要，这是设计师本身对生活的态度，对空间的理解。

1ST
2ND
3RD

LOVE WHAT YOU DO
DO WHAT YOU LOVE

EVERY THING WILL BE OK

If the style of furniture can embody the feature of space, then the combination of colors can enrich the expression of space, and lights can animate the atmosphere of space, the design of which just verifies these above points. It is a house with three bedrooms, which covers an area of 99 square meters, so it pursues spatial and functional practicality and perfectness all the time, boldly changes the graphic design of the original building, takes full advantage of architectural space's ductile or available position to conduct reasonable design and reuse. Thereinto, the space where the balcony in drawing room has been remodeled and designed includes the storage area, leisure area and working area. On the design modeling, designers integrate the modern simple design method into furniture and soft decoration with the retro style. On the combination of colors, they boldly use the strong contrast of colors to enrich space's sense of depth and keep space's young and positive vitality in the vintage elegance. At the same time, based on the pragmatic approach first, the design of light makes spatial brightness and depth more plump. Making big use of small space, design is not only a decoration but a need for life, which is designers' attitude to life and understanding on space.

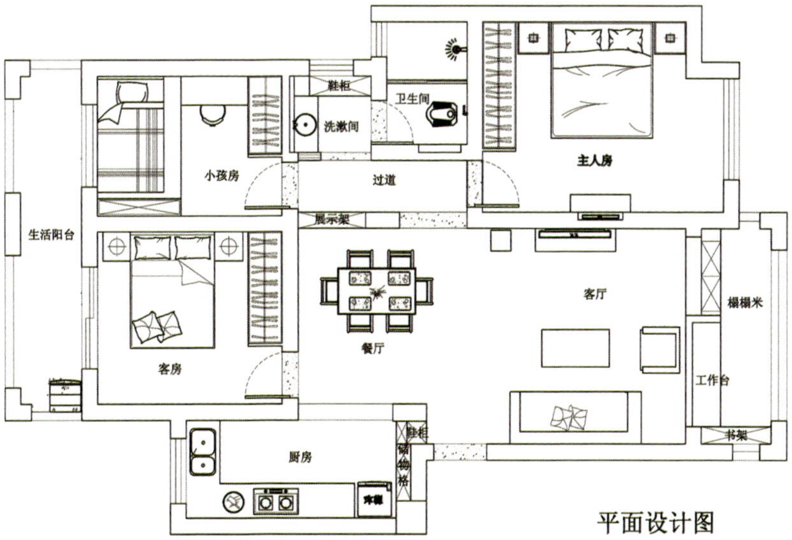

平面设计图

穗丰金湾样板房系列

SUIFENG JINWAN MODEL HOUSE SERIES

项目名称：穗丰金湾样板房系列
项目设计：广东星艺装饰集团
项目地址：广西北海
设计师：范建国　张郑承　许舰

本案为现代欧式风格，它沿袭古典欧式风格的主元素，融入了现代的生活元素。欧式的居室有的不只是豪华大气，更多的是惬意和浪漫。通过完美的曲线、精益求精的细节处理，反映出后工业时代个性化的美学观点和文化品位。

This scenario is Modern European style, it follows the main elements of classic European style and integrates modern life elements. Besides luxurious, European living rooms are more comfortable and romantic. It reflects personalized aesthetic view point and cultural taste in the post-industrial age by perfect curve and excelsior dealing on details.

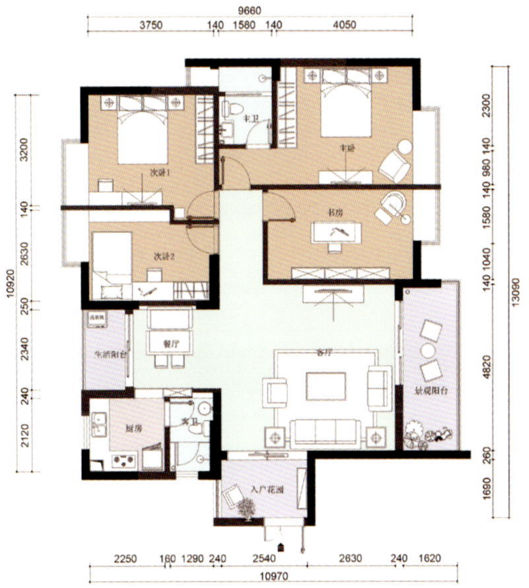

本案为美式乡村风格，它摒弃了烦琐和奢华，并将不同风格中的优秀元素交汇融合，以舒适机能为导向，强调“回归自然”，使这种风格变得更加轻松、舒适。

The scenario is Rustic American style, it abandons tediousness and luxury and fuses excellent elements in different styles. It is oriented by comfortable function and emphasizes “return to nature”, making the style more easier and cozier.

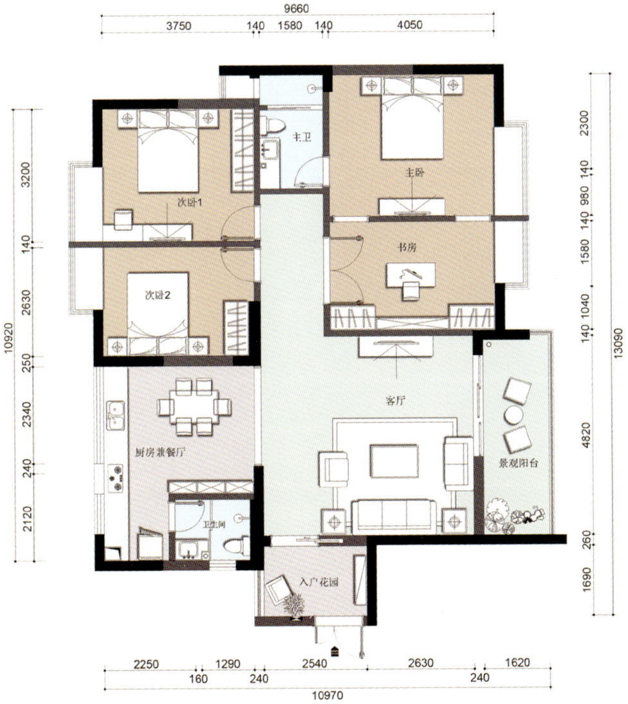
9660
3750
140
1580
140
4050
主卫
主卧
次卧1
书房
次卧2
客厅
厨房兼餐厅
景观阳台
卫生间
入户花园
3200
140
2630
250
2340
240
2120
10920
2300
140
980
140
1580
1040
140
4820
260
1690
13090
2250
1290
2540
2630
1620
160
240
240
10970

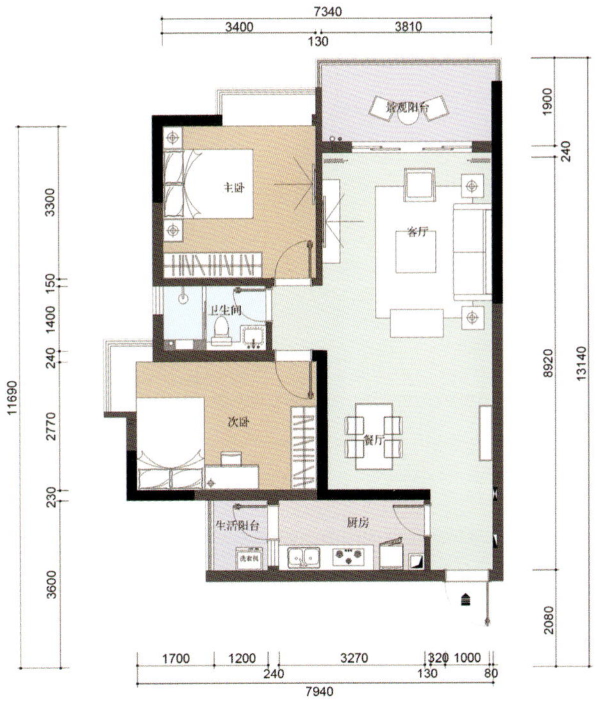

本案为新装饰主义风格，有别于传统古典主义的华丽感，它着重于实用，并在简约基础上彰显典雅的特性。在呈现精简线条的同时又蕴含奢华感，通过不同材质的搭配，展示人性化的舒适空间。

The scenario is New ornamentalism, different from traditional classical sense of luxuriance, it lays emphasis on utility and demonstrates the graceful characteristic based on the simplicity. It presents streamlined lines, contains the sense of luxury and shows the humanized comfortable space by the combination of different materials.

本案为“低调奢华”风格，是一种对传统奢侈设计概念的反思。简洁的室内设计虽然显得低调，装点室内空间的高科技材料和元素却能够在其中完全体现本身的品质和质感。在设计上，经典与现代元素被完美地结合起来，用简单的手法表现华丽气质，设计出更高品质的优质生活空间。

The scenario is Low-key and extravagant style, it is the reflection to traditional luxuriant design conception. Although the terse interior design looks low-profile, high-tech materials and elements decorating interior space fully reflect its own feature and texture. On design, classic and modern elements are combined perfectly. It employs simple technique to express the luxuriant temperament and designs a higher-quality living space.

本案为简约主义风格，将设计的元素、色彩、照明、原材料尽量简化，但对色彩、材料的质感加以着重表现，达到以少胜多、以简胜繁的效果，让居住者在这日趋繁忙的生活中，得到一个能彻底放松、以简洁和纯净来调节转换精神的空间。

The scenario is Minimalism, it tries to simplify the design's element, color, lighting and raw materials to the greatest extend, but it gives emphatic expression to the texture of color and material to achieve the effect that using the few to defeat the many and using the simple to defeat the complicated, which can make the inhabitants to get space where they are able to relax completely and adjust the spirit through simplicity and purity in the more bustling life.

本案为新东方主义风格，它是中式传统与现代元素的巧妙兼容，是中国传统文化意义在当前时代背景下的重新演绎。它将现代元素和传统元素结合在一起，以现代人的审美需求来打造富有传统韵味的空间。

The scenario is Neo-Orientalism, it is the artful compatibility between the traditional Chinese elements and modern elements, and the replay of traditional cultural Chinese significance under the current time background. It combines modern and traditional elements, creates the space full of traditional lingering charm with modern people's aesthetic needs.

长岛八百里香江

LONG ISLAND EIGHT-HUNDRED-MILES XIANG RIVER

项目名称：长岛八百里香江
项目设计：广东星艺装饰集团
项目地址：广西南宁
设计师：张旭

本案设计以不依张扬的美化、繁复的藻饰和不事雕琢的现代手法营构了一处欧洲贵族绅士内敛与精致的空间。雅士白与黑檀木的黑白组合在提升空间对比度的同时更带来了贵族的优雅简约气息。整个空间构图大气雍容，设计风格古典而有韵律。餐厅正中央，摆放着一处六人大圆桌。敞开式厨房，使你无论站在客厅、餐厅、厨房以及楼梯四者中任一处，视线都一览无遗，同时融进了更开阔的室外江景景观，也拉近了人与自然景观的距离。以美式田园设计元素为主体的二楼茶室，于角落处放置一棵仿真绿树，配合室外露台绿化，成就了一幅和谐画面，仿佛置身室外，临近大自然。窗纱随着微风轻摆，花香及茶香飘进室内，双门通道连接敞开式书房，让学习的心情更加轻松。

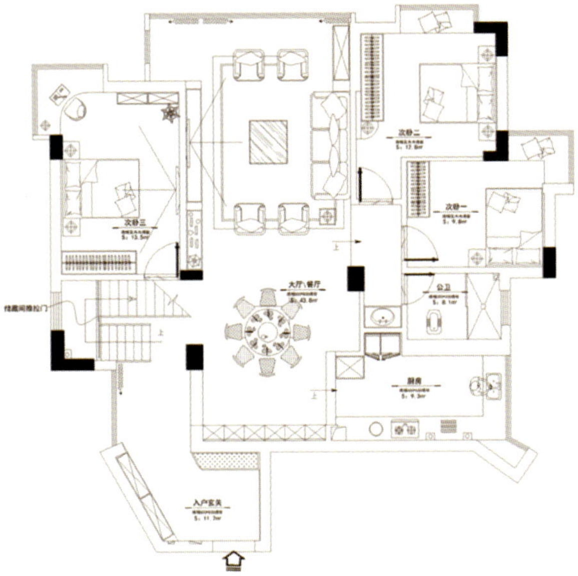

This scenario creates a reserved and exquisite space with the feature of European aristocratic gentlemen by the low-key beautify, complicated embellishments and natural modern techniques. The combination of Ariston and ebony not only lifts the space's contrast ratio but brings aristocratic elegant and simple feature. The whole space composition has the elegant atmosphere and the design style is classic and rhythmic. The midpoint of dining room puts a six-people large round table. The open-type kitchen makes you can get everything at a glance wherever you stand in one of there four places: drawing room, dining room, kitchen or stair, meanwhile, it integrates more open outdoor river landscape and make the distance between people and scenery closer. The tearoom in the second floor with the American rural design elements as the main body has an artificial green tree in the corner, matching with the outdoor terrace greening. They form a harmonious picture, which make people seem to stay outdoor and close to the nature. Window screens wave with the breeze, the fragrance of flowers and tea drift into the interior, the open-type study connects with two-door channel, all of which make people have a more relaxed mood to study.

龙胤花园

LONGYIN PARK

项目名称：龙胤花园
项目设计：广东星艺装饰集团
项目地址：广西南宁
设计师：范建国

本案试图诠释出新古典的美式格调。整个设计以更加温馨的手法营造出引人入胜的视觉效果，使人与空间有着更多的互动，设计当中的客厅与餐厅就是如此展现的。

This scenario tries to interpret the Neo-Classical American style. The whole design creates fascinating visual effect in warmer technique to make people and the space interact with each other. The drawing room and dining room in the design just display in this way.

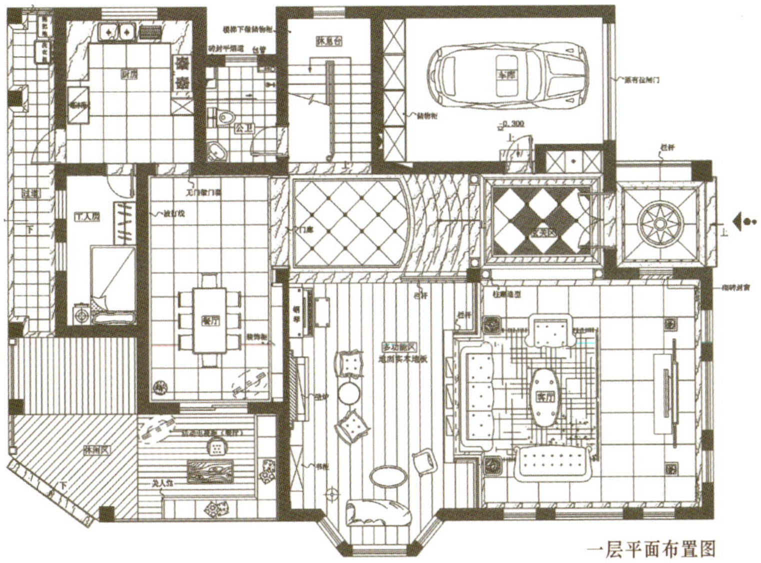

一层平面布置图

主卧室的设计给人一种尊崇感，镜面镂花的元素呈现出的是一份既现代又古典的气韵。空间的动线设计左右分流，让人走在空间中可以更加随意而自然，不至于受过多的干扰，当中体验不言而喻。

家具及软装陈设方面采用混搭的手法。明式圈椅与沉稳的美式家具很好地融合在一起，既满足了客户对美式田园生活的向往，又传承了中国的传统文化。

The design of master bedroom gives people a sense of respect. The specular engraving elements presents both a modern and classic artistic conception. The generatrix design about shunt in space makes people can walk in it more randomly and naturally and needn't be disturbed too much. The experience goes without saying.

On the furniture and soft display, it adopts the mixed technique. Round-backed armchairs in the form of Ming Dynasty and low-profile American furniture are integrated well, which not only satisfy clients' yearn for rural American life but inherit traditional Chinese culture.

彰泰·睿城

ZHANGTAI • RUI CITY

项目名称：彰泰 • 睿城
项目设计：广东星艺装饰集团
项目地址：广西桂林
设计师：夏少林　赵文强

本案设计师在中国风设计的基础上结合了时尚高贵和雅致的表现手法，打造出一个新东方生活风格。东南亚的塔顶天花板在合理装饰结构的同时突出了空间的特殊层次感，原木色体现了新东方的淡雅内涵。这些装饰与柔和的墙面和白色花格相结合，不仅让空间更舒适，而且让它看起来更宽敞；没有复杂的元素，只在关键处点缀上传统基础线条，通过现代材料和方法体现了东方传统文化。

Based on the Chinese-style design, the designer combines the noble and elegant techniques of expression to create a new oriental life style. The ceiling of the tower top in Southeastern Asia not only reasonably decorates structure but highlights space's special sense of depth. Burly-wood reflects new oriental simple and elegant connotation. These decorations combine with soft wall and white flower grid, which not only makes space more comfortable but makes it looks more capacious. It has no complex elements and only intersperse traditional based lines with key points. It reflects oriental traditional culture through modern materials and ways.

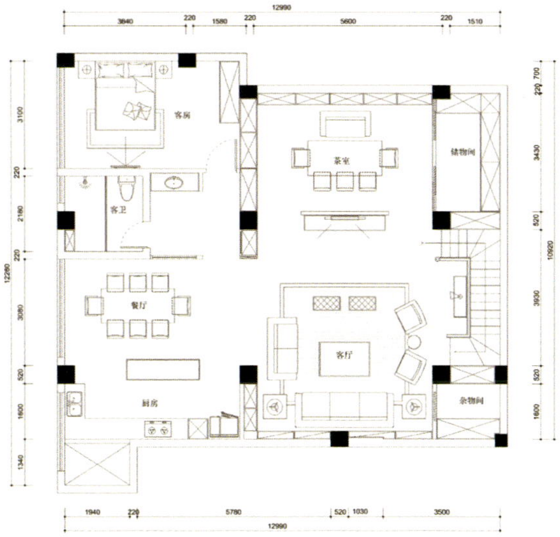

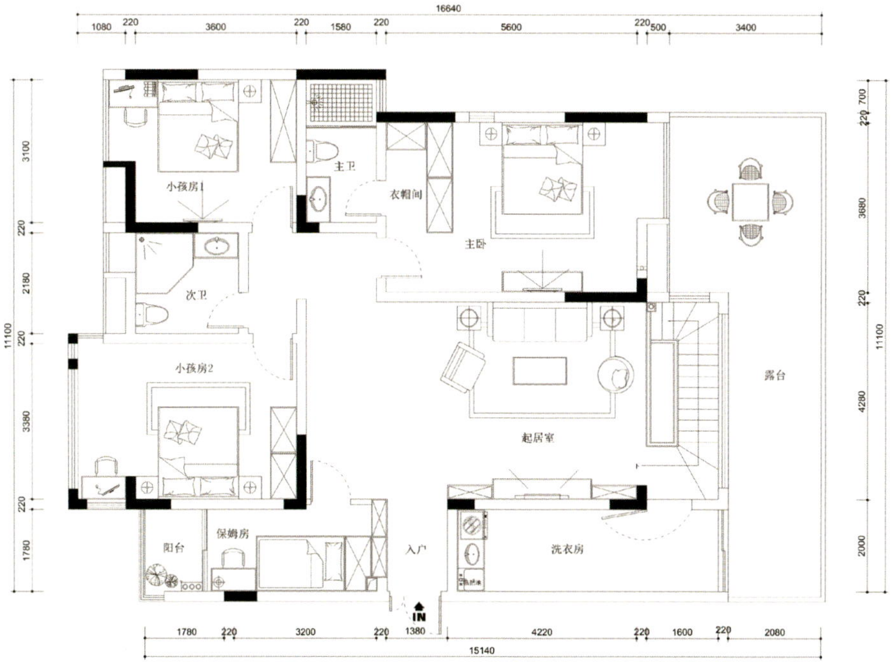
16640
1080
220
3600
220
1580
220
5600
220
500
3400
小孩房1
主卫
衣帽间
主卧
次卫
小孩房2
起居室
露台
阳台
保姆房
入户
洗衣房
IN
1780
220
3200
220
1380
4220
220
1600
220
2080
15140
3100
2180
3380
1780
11100
700
3680
4280
2000

耕艺种德

设计幸福